The Evolution of Meteorology

The Evolution of Meteorology

A Look into the Past, Present, and Future of Weather Forecasting

Kevin Anthony Teague

Nicole Gallicchio
Forecasting Consultants L.L.C

WILEY Blackwell

This edition first published 2017
© 2017 John Wiley & Sons Ltd

The right of Kevin A Teague and Nicole Gallicchio to be identified as the authors of this work has been asserted in accordance with law.

Registered Offices
John Wiley & Sons, Inc., 111 River Street, Hoboken, NJ 07030, USA
John Wiley & Sons Ltd, The Atrium, Southern Gate, Chichester, West Sussex, PO19 8SQ, UK

Editorial Office
9600 Garsington Road, Oxford, OX4 2DQ, UK

For details of our global editorial offices, customer services, and more information about Wiley products visit us at www.wiley.com.

Wiley also publishes its books in a variety of electronic formats and by print-on-demand. Some content that appears in standard print versions of this book may not be available in other formats.

Library of Congress Cataloging-in-Publication data applied for

ISBN - 9781119136149

Cover Design: Wiley
Cover Images: (Satellite) © Andrey Armyagov/Shutterstock; (Hurricane) © Harvepino/Shutterstock; (Barometer) © Mtsaride/Shutterstock; (Sun Flare) © solarseven/Shutterstock

Set in 10/12pt Warnock by SPi Global, Pondicherry, India

10 9 8 7 6 5 4 3 2 1

Contents

About the Authors

Two authors are responsible for the research and writing of this text: Mr. Kevin Anthony Teague, and Ms. Nicole Gallicchio. Both authors are BS Atmospheric Science degree holders, from the State University of New York, Stony Brook University.

Mr. Teague graduated from Stony Brook University in 2011. Along with his degree in Atmospheric Sciences, he also holds a BA in Psychology, attained from St. Joseph's College, New York. Together with these two degrees, Mr. Teague has used his unique and well-rounded educational background to research, analyze, and interpret meteorological processes in a very different way. The ability of Mr. Teague to then convey his knowledge in a professional yet extremely understandable manner has been what sets him apart.

Ms. Nicole Gallicchio graduated from Stony Brook University in 2012. Ms. Gallicchio has a strong education in programming, higher mathematics, as well as a degree in Atmospheric Sciences. She takes great pride in thoroughly understanding the math and science behind weather phenomena. Ms. Gallicchio has the ability to present her knowledge and expertise in a concise and quantitative way.

Together, Ms. Gallicchio and Mr. Teague founded Forecasting Consultants LLC. Forecasting Consultants LLC is a private weather forecasting and forensic company, geared toward a multitude of industries and users. Through their business, it has allowed the authors to grow and expand their knowledge base in all aspects of atmospheric science.

Preface

Meteorology, the study of weather, is a science that has been around since antiquity. Meteorology is a science that encompasses weather forecasting, climatology, atmospheric chemistry and physics, and oceanic interactions. These sciences are still being explored in great depth even though they have been studied for thousands of years. Advancements in the sciences have rapidly expanded in more recent years and are projected to continue to grow exponentially. Considering all the variables and unknowns in the atmosphere, meteorologists have become extremely accurate, with forecasts extending further out in time than ever thought possible. These forecasts are conveyed across the world and are reliable sources for many outlets. Throughout the world many different industries use weather forecasting for numerous tools and economic advances. The capabilities and possibilities that meteorology assists with far surpass many expectations, with updates to weather models, satellites, radars, and more constantly being sought after and put into use. This expansion is a relatively recent boom. Prior to computers and satellite technologies, weather forecasting was more guesswork than anything. Right up until the mid-twentieth century, meteorology was imprecise, often displayed in comedic ways on television and radio. Today's meteorology is nothing like 100 years ago, and that meteorology was nothing like 1000 years ago, and so on. Meteorology has gone from what was considered common knowledge that the moon and stars were signals of weather to come, and that weather was the forces of the gods, to what it is today, where weather is a quantifiable science, predicted and forecast days to months in advance with the help of computers that operate faster than the human brain. As knowledge grew and the field of meteorology was embraced, and as technological capabilities and worldwide cooperation increased, weather forecasting evolved to a daily "need" in the lives of all humankind.

This book was written to show the evolution of the field of meteorology, from its infancy in 3000 BC, through the birth of new ideas and the actual naming of the field as a science, and the technology boom, to today. This was not the original plan for this book. Originally, the authors were asked to write about extreme weather, and the scientific ideas and technology surrounding it. This subject matter cannot be fully understood without understanding of where the science was and where it is today. Upon researching the history of meteorology, it became very clear that there are very few concise single texts that show an outline of the history of the field of meteorology. The lack of resources out there led to the expanding of this text in order to include a full story of where meteorology was then to where it is now, and then not stopping there, but going into where the field is heading and what needs to be done to get the field to levels never before imagined. This book ends up being a comprehensive view of the history, but also a comprehensive view of forecasting technologies, organizations, governmental agencies, and world cooperative projects and legislation, along with a section dedicated to climate change theories and understanding, as well as extreme weather statistics and histories.

There are growing technologies in forecasting and public alerting when it comes to extreme weather. This is partially due to the increase in public knowledge of the devastation and possible threats that extreme weather poses. The cooperation between nations and companies in order to expand the reach and effectiveness of forecasts and planning when it comes to extreme weather is growing. There is a significant need to better prepare and combat weather's potential fury and devastation. It cannot be more evident that extreme weather is a global issue, not sparing any corner of the world from its possible wrath, coming in endlessly different ways – dust, wind, heat, cold, rain, snow, ocean depth and temperature, fires, floods, etc. No one area is protected from weather and no location is protected from the threat of extreme weather. This has always been the case, and this will always be the case. What has changed is the knowledge of extreme weather events, and the technologies, theories, inventions, and education in the field to help keep life, property, and livelihoods safe and out of harm's way as much as possible.

This book consists of five sections, starting with the ancient history of the first attempts to understand and predict weather. The book then flows into the very early birth of television, computers, and technologies useful to meteorology. The middle portion of the text consists of modern-day technologies, up through a few years ago. At this point, the text changes from a history book to a book of research, statistics, future paths, ideas, and suggestions, with a section on climate change and extreme weather events of today and beyond. There is also a section on the overall future direction of the field of meteorology. This text paints a broad picture of many ideas, each of which could have its very own 200-page text. It is created to help continue the debates, the expansion of ideas and knowledge, and to bring awareness to the overall path of the field of meteorology and where we may end up. The most fascinating aspect of this text is that it starts with the Ancient Babylonians and ends with the largest global agreement of any kind – the Paris Agreement.

Global research and scientific knowledge are at the forefront of expanding our understanding of meteorology and the atmospheric and oceanic sciences.

Acknowledgments

Writing this textbook was a complete honor, and we want to first off thank Justin Taberham for giving us the opportunity to be part of his series. After several weather and environmental conversations with Justin, we both seemed to have a professional connection that hopefully comes across in this text and throughout his series. We were delighted when we were first asked to be contributors to his series, and then when we were asked to author our own book we were overcome with excitement. We are both so very proud of the commitment and dedication to this project that we both showed and gave to each other.

Kevin: I greatly thank my wife Ashley for helping me cope with the stresses in all aspects of life, for helping me stay focused on all that is important, and for her never-ending love and commitment to me. Along with Ashley, I want to thank my parents, siblings, extended family, and friends, all of whom have shown an interest in this project. You have all helped me continue to work when I was hitting a wall, as well as helping me take a breather and decompress when I was getting too absorbed.

Nicole: I want to thank my immediate family for the support that they have provided me throughout this journey, along with the continued support in my future endeavors. My dad's relentless liveliness and my mother's kind words of encouragement have led me here today.

Throughout the research and writing of this book, we have been in touch with some of the greatest minds in the field of meteorology. With many conversations with staff from the National Weather Service, the Met Office, the World Meteorology Organization, and various agencies around the world, we thank you for your guidance, research, data, and all-round help in accomplishing our goals and tasks. We thank all the reviewers who took the time to read our proposal, give feedback and recommendations, and helped shape this book into what it is today. Your comments and feedback were of tremendous value. We want to thank Professor Brian Colle and Stony Brook University for providing us with a strong foundation. We also want to thank the various individuals who went above and beyond, helping us with permissions of use for images and research, especially, but not limited to: Christy Locke of UCAR, Susan Buchanan of the NWS, Rhys Gerholdt of the World Resources Institute, Markus Steuer and Jan Eichner of Munich-MR, Dean Lockett and Abdoulaye Harou of the WMO, Ken Mylne – Chair of OPAG(DPFS) for the WMO, Chantal Dunikowski of ECMWF, Clarke Rupert of the Delaware River Basin Commission, and Dmitry Nicolsky of the Geophysical Institute-University of Alaska Fairbanks. We also want to thank the very hard work of and great communication with Teresa Netzler at Wiley, as well as our original contacts at Wiley who have since moved on: Delia Sandford, Rachael Ballard, and Audrie Tan.

Through the work of our text and our business we look forward to continuing research and advancements in the realm of atmospheric science.

Section I

Building Blocks of Meteorology (3000 BC–AD 1950)

1

Ancient Civilizations, Philosophical Theories, and Folklore (3000 BC–AD 1400)

Kevin Anthony Teague

Understanding the history of meteorology and weather forecasting can only be accomplished just as any other subject: by learning the building blocks and uncovering the pioneers and the technological advancements of the field. Meteorology is the study of the atmosphere and its phenomena. What is known today about weather forecasting and meteorology is tremendously more comprehensive and complex than what was understood during ancient times, but at the same time is lacking and insufficient when looking at what is possible in the future. To first understand where we are today and where we may lead to in the coming years, we first have to look at where we came from. Although full sections or full texts could be dedicated to each of the following topics covered in this section – as well as many of the examples not even included – a solid foundation for the beginning of meteorology will be discussed, starting with one of earth's first-known civilizations and advancing to the brink of the first meteorological inventions.

1.1 Ancient Babylonians

Some of earth's earliest-known civilizations in ancient Mesopotamia first began record keeping and attempting to understand the skies dating back to 3000 BC. Calendars were created based on the lunar cycle, eventually leading to the development of a 360-day calendar, which was later adjusted to match the seasons based more on the solar cycle by King Hammurabi (1792–1750 BC) of Babylon. This was one of history's first acknowledgments of seasons and weather. Advancements continued, all based on the sun, moon, and stars. By the eighth century BC the ancient Babylonians, under King Nabû-nāṣir (747–734 BC), began to regularly keep records of the observations that were made of the sky. While mostly involving astronomy, these were some of the first recorded observations in history, and included time intervals of the rising and setting of the sun, eclipses, and position of the moon. To the Babylonians, much of what was observed was done by the supernatural; even so, they used these observations to begin to calculate when certain celestial events would occur in the future (Teresi, 2002, pp. 115–123).

A common technique used in their predictions was observing the halo around the moon (and sun). Certain halo conditions signified to the Babylonians that specific weather events, as well as other phenomena, would occur. One example was a report by King Nabû-šuma-iškun, (mid-eighth century BC) that "a halo (in the area) of Virgo portends rain and flood in winter," and a second states, "if the moon is surrounded by a black halo: the month holds rain, variant: clouds will be gathered" (Verderame, 2014, p. 100). These relate astrology to meteorology, and are a common theme throughout the Ancient Babylonian civilization. It was this type of data collection and observation that led the Babylonians to create some of the earliest versions of calendars and time keeping, as well as being among the earliest to predict and anticipate weather and natural events that could impact their world.

1.2 Ancient Chinese

The Chinese were believed to have begun keeping minimal records based on actual weather, including temperature, dating back to 1400 BC. Around 1216 BC, rainfall was recorded, as well as type of precipitation, whether it was sleet, rain, or snow, along with wind and wind direction. While the Chinese also believed that the supernatural was responsible for the types of weather conditions that occurred, such as bad weather meaning the gods were angry, the Chinese were still able to begin to understand the hydrological cycle. During the Han dynasty, (206 BC–AD 220), the understanding of the hydrological cycle went from the idea of qi, or the energy of the earth and heavens, to the understanding that clouds were created from the rising waters due to evaporation. By the ninth century AD, the Chinese figured out a way to measure the amount of moisture in the air, by weighing charcoal as it absorbs moisture in the air. The Chinese also began to understand tides and past floods by locating water fossils on high land areas not commonly associated with water (Teresi, 2002, pp. 246–247, 257).

As the Chinese developed socially, their version of meteorology did as well. One example was the developing Chinese belief in the forces of yin and yang, which need to be in balance for the world, health, and morality to be in agreement. The yin and yang have tremendous influence, even today, especially in areas such as health and medicine. The Chinese also related the yin and yang to weather, needing perfect balance of the yin (clouds and rain – earth-bound elements) with the yang (fire and heat of the sun – celestial elements). Certain weather events were believed to have occurred because one of the elements was more dominant than the others at those specific times (Hamblyn, 2001, p. 23).

The Chinese based their calendars on the moon, sun, and the weather dating back to the Qin (221–206 BC) and Han dynasties. The Chinese almanac still acknowledges this thinking, and divides the year up into 24 seasons, or festivals, each about 15 days long. These seasons mark weather events that typically took place during specific parts of the year. From "The Beginning of Spring" to "The Waking of the Insects," "Corn Rain," and "Great Cold," each season marked their own effects on the Chinese agricultural lifestyle (LaFleur, 2010, pp. 425–426). These seasons, at the time, showed an accurate interpretation of the weather that occurred or was to occur, and each season can be viewed as an early style of climatology and weather forecasting.

1.3 Aristotle and *Meteorologica*

Perhaps the most influential early contributor to the beginning of meteorology is the Ancient Greek philosopher Aristotle, 384–322 BC. Aristotle is universally thought to be one of the world's greatest thinkers. One of his most important pieces of work was his treatise *Meteorologica*. While this work includes chapters on astronomy, water, physics, and more, *Meteorologica* has very important sections on weather. He studied meteorology as part of his investigation of physics. Physics, to Aristotle, was the equivalent to what would now be called natural philosophy, or the study of nature (Hollar, 2013, p. 14).

In a sense, some of Aristotle's views involved the supernatural, and didn't always go into root causes of some phenomena. Instead, Aristotle stated that there were four bodies – fire, air, earth, and water – that related to four principles – hot, cold, dry, and moist. Aristotle held fire as the highest body and earth as the lowest, and stated that they were the "material causes of the events in this world" (Webster, 1955, p. 339a). He took this supernatural viewpoint and then put his own spin on what he observed. Aristotle based a lot of his viewpoints on the sun, going as far as saying the reason the night is calmer than day is due to the absence of the sun.

Inside *Meteorologica*, Aristotle acknowledges the understanding of warm air rising and cooler air falling. This relates greatly to today's understanding of warm air being less dense and rising and expanding, while cooler air is more dense, and is therefore falling and condensing. This idea of hot and cold is connected also with Aristotle's early interpretation of a lower and upper level of the atmosphere, lower being the surface and upper where the clouds are located. He also describes precipitation as being broken down into three types: rain, snow, and hail. These precipitation types are then correlated to the upper or lower atmosphere. He explains that snow and frost are one and the same, the only difference being snow is in the upper region and is when clouds freeze, and frost is in the lower region and is when vapors freeze. He states that rain and dew are the same – the only difference being that rain occurs in the upper region and is in greater quantity than dew, which occurs in the lower region. The third, hail, Aristotle explains, is ice that forms mainly in the spring and autumn in warmer locations, and occurs only in the upper region, but very close to the surface and not high in the clouds. His reasoning for this was based on the fact that hail is often large, jagged, and non-circular. Aristotle felt that if hail were to have fallen from high up in the upper region the colliding stones would have broken themselves up into round and smaller stones and not the large angular ones that occur (Webster, 1955, pp. 346b–348b).

As for winds, Aristotle states that north winds are cold and south are warm. He states that east winds are warmer than west winds due to the origin of the sun in the east. Aristotle even goes as far as to combine the west winds with the north and east winds with the south. This shows that the basis of Aristotle's understanding of wind heavily relies on theories revolving around hot and cold. Aristotle's depth of wind includes theories on hurricanes as well, explaining that hurricanes are created by the converging of winds, which then begin to veer around and fall in on each other, which is his explanation for the eye of a hurricane (Webster, 1955, pp. 362b–365a).

Some of the more glaring errors in Aristotle's work include that it snows and rains more at night than in the day, along with his usage of the words "always" and "never" to describe meteorology. Weather is not an "always" or "never" type of science, and many variables contribute to the final outcome, with each weather event different from the one before. In *Meteorologica*, Aristotle takes a more concrete viewpoint, seeing things as certainties and very cut and dry, albeit with often vague explanations.

Aristotle's observations were new to the Greeks and his knowledge and breakthroughs were beyond anything of its time. Even though Aristotle's interpretations of weather and the atmosphere were not extremely accurate compared to today's standards, *Meteorologica* was the main go-to theory on weather for the next 2000 years. Aristotle's influence sticks with us even today, as he wasn't just the pioneer of the science of meteorology but also the word "meteorology" itself (Webster, 1955, pp. 338a–390b).

1.4 Theophrastus and the *Book of Signs*

Theophrastus (371–287 BC), whom Aristotle named as his successor at the Lyceum, is often credited for writing the treatise *Book of Signs*. Most of Theophrastus' work is based on lore type ideas (discussed further in Section 1.7), while other work is based on no real factual backing, such as his theory that shooting stars are a sign of impending rain and wind and will come into an area from the direction of the corner of the sky that the shooting stars were occurring (Thayer, 2012, p. 399). He touched on a type of climatology when he stated that an abundance of winter rains indicates a dry spring, and vice versa (Thayer, 2012, p. 407). Theophrastus' interpretation regarding wind includes theories that a northerly wind will occur when the top part of a crescent moon is bent, and a southerly wind when the bottom of the crescent moon is bent (Thayer, 2012, p. 407). Also, winter will be calm in regards to wind if there is a long period of time when there are periodic winds as well as a windy autumn season leading up to the winter (Thayer, 2012, p. 415). Many other similar types of weather explanations are included in the *Book of Signs* and were never grasped as solidly as Aristotle's *Meteorologica*, as they were based more on the lore of the time and were very hard to comprehend.

1.5 Ibn Wahshiyya and *Nabataean Agriculture*

These non-scientific approaches to the understanding of weather and weather forecasting continued into the tenth century AD. This was demonstrated by the work of an Iraqi named Ibn Wahshiyya. He is credited for translating the *Nabataean Agriculture*, which has many mysteries surrounding it, including when and where it was originally written (anywhere from about AD 100–900), who wrote it, and what its true meaning is. Even with these mysteries, the ideas of weather and weather forecasting occur throughout the text, due to their importance and influence on agricultural development. Wahshiyya translated a section that shows the significance placed on the moon. One translation includes an example that seeing the moon at night indicates clear weather ahead, and the longer you observe the moon, the longer the clear weather will persist. He explains that a halo around the moon also indicates clear weather to come. Similar examples are seen

when the sun is observed as well. Wahshiyya writes about the connection between cloud coverage and weather, with examples such as low-lying clouds that dissolve away quickly being an indication of cold weather, and warm weather indicated by clouds that grow larger, fall apart into smaller clouds, and then dissolving altogether.

Also stated in the *Nabataean Agriculture* is the use of animals to help indicate weather, with one method being the hooting of an owl as an indicator for the ending of cold or bad weather. Other examples taken from Wahshiyya's translation involve wind predictions, such as a west wind's anticipation when a fourth night moon has a red-appearing crescent. North winds are also predicted by the occurrence of storms and lightning, such as when lightning is seen to the south while clear skies are observed overhead. This example also indicated that rain would come in from the south. One final example taken from this translation that shows the lack of understanding at the time is the indication of extreme cold, which Wahshiyya explains could be done when one sees sparks fly off clay pots when they are struck or scraped (Hameen-Anttila, 2006, p. 80–84). All these examples are more in line with the theories of Theophrastus, more superficial with less actual advancements to the understanding of weather, and based more strictly on observations from the past and lore-like theory.

1.6 William of Ockham and Nominalism

Nominalism regarded the universal idea or problem of universals as unreal. The problem of universals questions the existence of the properties or category of a particular phenomenon. Instead, nominalists regard only the concrete individual thing as real. This competed greatly for one of the first times against Aristotle. The work of William of Ockham (AD 1285–1347) used knowledge, understanding, and reasoning to start the beginning of a change in physics and natural sciences, while in the past Aristotle and others based the natural sciences more on assertions (Hirschberger, 1977, 65, 92–94). Ockham focused a lot on Aristotle's work, stating that the category of quantity is not needed and that mathematics is not real since it is a universal. His nominalist approach further states that mathematics needs to expand to substance and to qualities ("William Ockham", 2015), which in theory would take away the vague assertions and instead make things more concrete, or real. In other words, nominalism was the beginning of taking away all of the ideas and theories of the past that were subjective, such as Aristotle's theories in *Meteorologica*, and leaving only the objective, or seeing what is in front of you as exactly what it is. This frame of thinking lays the stage for the foundation behind modern science.

1.7 Religion, Folklore, and Animals

While the idea of an accurate concrete science behind weather knowledge was still lacking, theories began to emerge from the works of Aristotle, Wahshiyya, and Ockham, as stated previously. With that said, there has always been another important aspect to weather forecasting, dating all the way back to ancient history and that is with the use of religion and folklore. Babylonians had their beliefs of supernatural causes in connection with the celestial events that took place, and the Chinese believed that the angry gods produced dangerous weather. There were many other examples of greater beings controlling the weather, some into the eighth century AD and more recent, some of which are still practiced today.

Nearly 3000 years ago in the Wisconsin River area of North America, some of the earliest civilizations in that region prayed to the Green Water Spirit. One of those early civilizations was the Ho-Chunk tribe. Part of a long family line of warrior chiefs, Albert Yellow Thunder (1878–1951) explained that the Green Water Spirit was responsible for the melting of snow

and ice, and the creation of lakes and streams from the melted waters. He also was responsible for creating game for hunting and the trees that surrounded the area (Kavasch, 2004, pp. 8–9).

The Vikings in the eighth century AD believed that two specific gods were responsible for the weather. One was Thor, the god of thunder. Since the Vikings were sailors, they spent many days at sea. Their Scandinavian location was prone to strong storms over the waters of the North and Baltic seas that were attributed to the wrath of Thor. Thunder was the sound of Thor in his chariot, and lightning was the flashes of the sparks of Thor's ax killing his enemies. The second god was Freyr, the god of rain, sun, and fertile earth. Freyr was of huge importance because it was he who helped produce the Vikings plentiful crops (Reynolds *et al.*, 2008, p. 26).

The thirteenth century AD involved the Aztecs, who believed strongly in many deities, some of which included weather aspects, especially Tlaloc, the god who controlled the weather. The Aztecs were an agricultural civilization, relying heavily on fertile soil, and proper weather conditions to keep crops rich. The location of the Aztec civilization in Mexico is in an area that experiences both drought and flooding. It was believed that Tlaloc was the one who decided which would occur. The Aztecs believed so strongly in Tlaloc that they even sacrificed children in order to keep Tlaloc happy and their lands healthy (Reynolds *et al.*, 2008, p. 26).

More in line with the folklore aspect of beliefs, rain dancing and weather sayings were and still are used today. Rain dancing has been practiced by many cultures in order to bring on the rains during times of need. Cultures across the world still practice rain dancing (Reynolds *et al.*, 2008, p. 27). Just as the rain dancing continues to this day, wise weather sayings are still often heard spoken in everyday society. Some of these include: "red skies at night, sailor's delight" and "red skies in the morning, sailor's warning." These sayings have some basis in truth, even though they are not always right. They are based on the idea that when there is a red sky at night dry conditions are moving in, often as a high pressure system, as the sun's rays shine through dry particulate matter, thus causing the red coloring. Since weather typically moves from west to east, the idea of red skies at night in the sunsetting western sky means dry weather is moving in from the west. A red sky in the morning means the opposite: bad weather will be replacing the calm dry weather or that the high-pressure system creating the dry weather allowing the morning sun in the east to shine and become red in coloring will continue to move eastward and away. Some weather lore, based on patterns such as the example above, were first used as forecasting methods as far back as 300 BC, and while these do have some valid meanings behind them, they are of course not always accurate, but amazingly they are still leaned upon in today's society.

Besides folklore and religion, animal behavior is known to be a signal of natural phenomena to come, including weather. Using animal behavior to forewarn about impending weather is an ancient practise dating back to the Ancient Chinese, and later the Greeks in the *Book of Signs* by Theophrastus, and in the *Nabataean Agriculture* by Ibn Wahshiyya, as well as still being relied upon today by many. Be it observing dragonfly behavior, the sounds of frogs, the singing and flying patterns of birds, even anthill development and livestock behavior, many cultures and civilizations tied weather events to these observations and used them as their tool for weather forecasting (Ehlert, 2012, pp. 105–106).

Chapter Summary

These beliefs, and lore, along with the first established theories and approaches to understanding weather by the Babylonians, Chinese, Aristotle, and Wahshiyya, to name a few, were the building blocks of weather knowledge and meteorology. All that was accomplished and learned up to the fifteenth century was done mainly without the use of technological assistance. This all changed quickly, as did the understanding and future capability to forecast weather.

References

Book References

Ehlert, J. (2012) *Beautiful Floods: Environmental Knowledge and Agrarian Change in the Mekong Delta, Vietnam*, LIT Verlag Münster, Germany.

Hamblyn, R. (2001) *The Invention of Clouds: How an Amateur Meteorologist Forged the Language of the Skies*, Farrar, Straus, and Giroux, New York.

Hameen-Anttila, J. (2006) *The Last Pagans of Iraq: Ibn Wae Shiyya and his Nabatean Agriculture*, Koninklijke Brill NV, Leiden, The Netherlands.

Hirschberger, J. (1977) *A Short History of Western Philosophy*, Westview Press, Boulder, CO.

Hollar, S. (2013) *Inventors and Innovations: Pioneers in the World of Weather and Climatology*, Britannica Educational Publishing in Association with Rosen Educational Services, New York.

Kavasch, E.B. (2004) *The Mound Builders of Ancient North America: 4000 Years of American Indian Art, Science, Engineering, & Spirituality Reflected in Majestic Earthworks & Artifacts*, iUniverse, Inc, Lincoln, NB.

LaFleur, R.A. (2010) Chinese Festivals and National Holidays, in *Asia in Focus: China*, ABC-CLIO, LLC, Santa Barbara, CA.

Reynolds, R., Hammond, J.C., Smith, F., and Tempest, S. (2008) *Eyewitness Companions: Weather-Forecasting, Weather Phenomena, Climate Change, Meteorology*, (eds. R. Gilbert, D. John, R.S. Colson, *et al.*), Dorling Kindersley Publishing, New York.

Teresi, D. (2002) *Lost Discoveries: The Ancient Roots of Modern Science: From the Babylonians to the Maya*, Simon & Schuster, New York.

Verderame, L. (2014) The Halo of the Moon, in *Divination in the Ancient Near East*, (ed. J.C. Fincke), Eisenbraun, Winona Lake, IN.

Webster, E.W. (1955) Meteorologica, in *The Works of Aristotle*, vol. 3, (ed. W.D. Ross), Clarendon Press, Oxford.

Website References

Thayer, B. (updated 13 May 2012) "Theophrastus." *De signis by Theophrastus.* Published in Vol. II of the Loeb Classical Library edition of the *Enquiry into Plants*, 1926, http://penelope.uchicago.edu/Thayer/E/Roman/Texts/Theophrastus/De_signis*.html, accessed 4 April 2015.

"William Ockham." (2015) *Project Gutenberg Self Publishing Press.* World Public Library, http://self.gutenberg.org/articles/William_Ockham, accessed 4 April 2015.

2

Inventions of Weather Instruments (1400–1800)

Nicole Gallicchio

Throughout history the goal of humans who have studied, observed, recorded, and researched the weather has been accuracy and predictability. As beliefs and folklore began to take a back seat to scientific approaches, new inventions and technologies began to emerge. By the fifteenth and sixteenth centuries, new inventions aided the ability to understand and forecast weather. These new inventions became known as "weather instruments." Weather instrument ideas and prototypes were being developed all over the globe throughout this period. There were several successful scientists who provided us with the foundations for our current meteorological technologies and instrumentation.

2.1 Rain Gauge

Precipitation is any form of condensed atmospheric water vapor that falls to earth due to gravity. This precipitation can be in the form of rain, sleet, snow, dew, and hail, and can be measured and recorded. The rain gauge is a meteorological instrument that measures precipitation over a specific period. As precipitation falls it is collected in this device and recorded. The history of this instrument shows that it has been around since ancient times, with records of the simplest rain gauges going back as far as the sixth century BC in ancient Greece. In India around the fourth century BC passages in the book *The Arthashastra* by Kautilya state, "A bowl as wide as an Aratin [46 cm], shall be set up as a rain gauge" and "according as the rain fall is more or less, the superintendent shall sow the seeds, which require more or less water." From the second century BC to the second century AD there are scriptures written down from the Jewish oral traditions, "The Mishnah," about the amount of rainfall collected/measured over what is assumed to be a year (Strangeways, 2010, pp. 3–4). The first recorded snow gauge, similar premise as the rain gauge, was used in China in AD 1247.

There were few additional advancements to the precipitation gauges until the fifteenth century, when simple devices for measuring precipitation began to take shape. Korea used simple manual rain gauges around AD 1420 and continued to use this type throughout the sixteenth century. This device was a basic cylindrical metal cup that collected precipitation. In other parts of the world, Italian scientist Benedetto Castelli (1578–1643) was the first to construct the rain gauge in Europe in 1639. As technology began to evolve, further advancements and inventions were created. English physicist and instrument inventor Robert Hooke (1635–1703) contributed many inventions and ideas to meteorology. Robert Hooke

The Evolution of Meteorology: A Look into the Past, Present, and Future of Weather Forecasting, First Edition.
Kevin Anthony Teague and Nicole Gallicchio.

and Sir Christopher Wren, English scientist and mathematician, are both credited with the invention of the first automatic rain gauge. This new invention became known as a "tipping bucket rain gauge." Later, in May 1670, Wren was ordered to accept Hooke as the inventor of the device (Biswas, 1967, pp. 94–95). In May of 1679, it was recorded that Hooke had completed the instrument and had also completed creating his own manual rain gauge that consisted of a funnel and collecting globe at the bottom. An English mathematician and astronomer, Richard Towneley, recorded the first substantial rainfall measurements from 1677 to 1704 (Reynolds *et al.*, 2008, p. 35).

Developments of the rain gauge continued into the eighteenth and nineteenth centuries. In the late 1770s, a Dr. D. Dobson (1777) of Liverpool was one of the first to use a large flat open area in order to collect the rainfall. This idea led to experiments in raindrop size and evaporation (Strangeways, 2004, p. 1). Further experiments with raindrop size were conducted and concluded that rain gauges on the roof only caught 80% of rain when compared to one placed on the floor. This conclusion was based on the theory that drops increase in size over the fall to the ground (Strangeways, 2004, p. 15). A few years later the improvements on the rain gauge continued and William Jevons (1835–1882), an English economist and logician, corrected the theories and errors in the previous devices. Jevons explained that wind systematically creates error in the collections, correcting the previous theories of drop size change. George James Symons (1838–1900), a British meteorologist who founded the British Rainfall Organisation, helped complete the work of Jevons and expanded his theories. Symons is known as one of the most important scientists in the history of rainfall measurement. Since the late nineteenth century, there have been numerous modifications of the rain gauge. These additions to the basic idea of collecting and measuring precipitation have only increased accuracy over time.

2.2 Hygrometer

The amount of moisture in the air is called "the humidity." It is the element of the atmosphere that is most frequently changing. A hygrometer measures atmospheric humidity by determining the amount of water vapor in the air. Although the ancient Chinese attempted to use this concept of air moisture measuring, the earliest records of a more modern version of the hygrometer derive from the early fifteenth century. Around 1450, cardinal and mathematician Nicholas of Cusa created a plan for a device known as a hygroscopic hygrometer. His device detailed hanging wool on one end of a large scale with stones on the other end. This scale would be balanced and level when the air was temperate. If it was humid (increased moisture), the scale would tilt toward the wool. Wool is a very porous material and easily absorbs moisture. When the wool absorbs the moisture in humid weather, it becomes heavier and therefore tips the scale toward the wool side. The opposite would occur in dry air, as the scale would tilt toward the stones as the wool dried out, becoming lighter in weight. There has been no significant evidence found proving that Nicholas of Cusa created a prototype of his idea. However, in 1481, Leonardo de Vinci used this idea to create his own hygrometer. It was not until the mid-seventeenth century that the hygrometer made any further improvements or advancements.

As the latter part of the sixteenth century approached, different compositions of hygrometers began to emerge. Around 1650, Ferdinand II de' Medici (1610–1670), the Grand Duke of Tuscany, invented the condensation hygrometer ("In-Depth: Hygrometer", 2010). The condensation hygrometer works when water vapor comes in contact with cylindrical iced glass. The vapor then condenses, forming water droplets that descend into the graduated glass below. Shortly after, around 1660, Francesco Folli (1624–1685), a doctor of philosophy born in Poppi, invented a paper ribbon hygrometer, which he called *mostra umidaria* ("In-Depth: Hygrometer", 2010).

The paper acted as a hygroscopic substance and was strung across a wooden rod. This wooden rod acted as a dial as the length of the paper changed due to the humidity. Several years later, in 1687, French physicist Guillaume Amontons (1663–1705) made improvements with mercury-filled bags. In 1775, Horace-Bénédict de Saussure (1740–1799), a Swiss aristocrat and physicist, invented the first hair hygrometer, which is still around and widely used today (Allaby, 1998, p. 9). The hair hygrometer works by the changes in length of a strand of human hair. The hair changes length as the humidity increases or decreases. In 1820, John Frederic Daniell (1790–1845), an English chemist, invented the dew-point hygrometer (The Editors of Encyclopaedia Britannica, 2015). "This consists typically of a polished metal mirror that is cooled at a constant pressure and constant vapor content until moisture just begins to condense on it." This temperature at which condensation begins to form is referred to as the "dew point temperature."

2.3 Thermometer

Touching upon Aristotle's views, there are four bodies – fire, air, earth, and water – that relate to four principles – hot, cold, dry, and moist. The polarity of "hot" and "cold," just like that of "moist" and "dry," derives from our sense of touch. As stated in Section 1.3, Aristotle introduced the ideas of hot and cold, relating them to different atmospheric elements. He made no attempt, however, to put numerical quantities to his theories. Aelius Galenus (AD 129–216), Greek physician and philosopher, introduced the idea of "degrees of heat and cold," assigning a neutral point between ice and boiling water. This is the earliest approach to assigning a fixed mark or standard of temperature. The scale of temperature was prominently associated with physicians, even though they had no instrument to directly measure it with. The invention of an instrument that differentiated between hot and cold was first referred to as a "thermoscope."

The difference between a thermoscope and thermometer is that a thermometer provides a scale to the differentiation. As to the original invention of the thermometer, it remains negligible. As to the invention of the thermoscope, it becomes the question of when an experiment exponentiates into an invention.

The first steps toward the invention of the thermometer were taken by Philo of Byzantium (280–220 BC) and Heron of Alexandria (AD 10–70), both Greek engineers. Their works demonstrated the expansion and contraction of certain substances, notably air. The experiments/devices used the notion of a closed tube that was partially filled with air and had a container of water at the other end. The expansion and contraction of air caused the water/air line to change along the tube. These devices would later become known as thermoscopes. During the late sixteenth century, the works of these scientists were translated and published. There are four scientists that are known for the invention of the thermometer. They are considered in pairs, due to their geographic locations at the time of the accomplished work. Galileo Galilei and Santorio Santorio lived in Italy, while Cornelius Drebbel and Robert Fludd lived in the North America, north of the Appalachian Mountains. In 1594, Galileo had read the published works of Heron and notably enhanced the achievements of his predecessors. He is said to be the inventor of the thermometer; however, what he actually produced were thermoscopes that reflected changes in sensible heat. Galileo also discovered spherical tubes that were filled with aqueous alcohol at different densities that would rise and fall according to changes in temperature. Today this device is known as an "air thermometer" because it is calibrated and marked with appropriate temperature scales. Santorio Santorio (1561–1636), also known as Francesco Sagredo, is known as the first person to attach a scale to the thermoscope; creating the thermometer, in 1612.

Across the Appalachian Mountains, Cornelius Drebbel (1572–1633) was working on a device, the "Perpetuum mobile," that operated on the basis of changes in atmospheric temperature and pressure. He used his technology from this device to create what is known today as the "thermostat." The thermostat was a thermometer that regulated the temperature on ovens and furnaces. In 1638, Robert Fludd (1574–1637) was the first to publish a diagram of the thermometer, a thermoscope that contained a scale.

Each inventor and each thermometer had a unique scale, leading to a lack of uniformity. The development of a temperature scale occurred over a long period and by numerous scientists. It should be noted that quantities of temperature are not additive and that properties of different elements change with changes in temperature. Therefore, it was difficult to arrive at a universal basis temperature scale. In 1665, Christiaan Huygens (1629–1695), an astronomer and physicist, suggested using the melting and boiling points of water as standards, and in 1694, Carlo Rinaldini (1615–1698), a mathematician, philosopher, and Italian meteorologist, proposed using the melting and boiling points of water as fixed points on a universal scale. In 1701, Isaac Newton (1642–1727) published a systematic account of ranges in temperature, "Degrees of Heat," proposing a scale of 12 degrees between the melting point of ice and body temperature. Finally, in 1724, Daniel Gabriel Fahrenheit (1686–1736), a German physicist, created a temperature scale which famously is named after him. He created a thermometer using a mercury bulb, which has a high coefficient of expansion, and a capillary tube. He was able to refine the scale of temperature and create a consistent reproduction of his device (Fowler, 2008). In 1742, Anders Celsius (1701–1744), a Swedish astronomer, physicist, and mathematician, created a scale in which 0 represented the boiling point of water and 100 designated the freezing point of water. In his paper he remarked on the effects that pressure has on the melting and boiling point of water. He proposed calibrating temperature with mean barometric pressure, today known as "one standard atmosphere" (atm). In 1744, Carolus Linnaeus (1707–1778), Swedish botanist, physician, and zoologist,

reversed the Celsius scale, which is its current known measurements, creating 100 degrees as the boiling point of water and 0 degrees as the freezing point. The history of the thermometer scale cannot be completed without the discussion of the Kelvin scale. The Kelvin scale was developed by William Thomson (1824–1907), the first Baron Kelvin, British mathematician and physicist. The idea was discovered in the nineteenth century as the relationship between volume and temperature of gas was researched. Scientists theorized that the volume of gas should become zero as the temperature became −273.15 degrees Celsius. In 1848, Thomson used this in order to create an absolute zero temperature scale. Absolute zero is based upon the idea of a temperature in which molecules cease all movement. Thomson used the degrees in Celsius to calculate his increments (Zimmerman, 2013). The Kelvin scale has become very useful in the scientific community because of its lack of negative numbers. The thermometer is arguably the most prominent and influential meteorological instrument in history. Today, there is a significant number of different types of advanced thermometers that all stem from the premises of our scientific ancestors.

2.3.1 Psychrometer

With the inventions, ideas, and advancements of the hygrometer and thermometer, the theories of combining the two began to emerge during the sixteenth and seventeenth centuries. In 1818, a German inventor, Ernst Ferdinand August (1795–1870), patented the term "psychrometer," deriving from the Greek language meaning "cold measure." The psychrometer is a hygrometric instrument based on the principle that dry air enhances evaporation, opposed to wet air, which hinders it. The instrument works by having two identical thermometers side by side, one thermometer is equipped with a dry bulb and the other with a wet bulb. The wet bulb thermometer is kept continuously moist by a wet band. As the atmospheric humidity changes, the temperatures will change accordingly. If the air is dry, the rate of water evaporating on the wet bulb increases, which in turn lowers the temperature, due to evaporation processes, of the wet bulb thermometer. The temperatures are read from both thermometers and the difference is correlated with tables, and the humidity can be calculated. There will be no temperature difference when the relative humidity is 100%. Understanding the properties of the atmosphere allows for greater accuracy in weather forecasting. Combinations of weather instruments continued to flourish as time went on.

2.4 Barometer

The atmosphere consists of a layer of gas that is retained by earth's gravity. This layer of gas is commonly referred to as "air." The weight of the air is called atmospheric or barometric pressure. Pressure in a fluid is defined as the force per unit area exerted on any flat surface. In the case of earth's atmosphere, there are no constraints to its volume. The pressure exerted at any level of the atmosphere is almost solely based upon the weight of the air pressing down from above, resulting from gravitational attraction. Pressure in one specific place is always changing, and variations in horizontal pressure are much less than those of vertical pressure. The ideas first considered in antiquity were the basis for the invention of the barometer. The idea surrounding the concept of a vacuum space can be considered the premise of the barometer. There was great debate at the time as to whether a vacuum could exist.

There have been many scientists over time that have made contributions to the realms of physics and chemistry which has led to the development of the barometer. The Italian physicist and mathematician Evangelista Torricelli (1608–1647), who can loosely be called a pupil of Galileo's, is generally credited with inventing the barometer. Torricelli's experiment began by testing theories on air pressure and vacuum. He tested his theories by using a glass tube about a meter long sealed at one end and filled with mercury. He then inverted the tube into a dish filled with mercury. Some of the mercury drained out of the glass tube and into the dish. Torricelli explained this by "assuming that mercury drains from the glass tube until the force of the column of mercury pushing down on the inside of the tube exactly balances the force of the atmosphere pushing down on the surface of the liquid outside the tube" ("Barometer (Evangelista Torricelli)", n.d.). The height of the mercury column changed from day to day, as Torricelli predicted, coinciding with the changes in atmospheric pressure. His invention of the barometer, around 1643, led to numerous physics, chemistry, and atmospheric advancements. To dedicate and recognize his achievements, the pressure unit known as "torr" was named after him. The apparatus of a barometer, however, was not made by Torricelli himself, because he had died, but by his friend Vincenzo Viviani (1622–1703), whom he left his experiments to.

During the seventeenth century, there were several types of experiments made to construct different barometers consisting of various liquids. No fluid seems to give a more accurate reading of pressure than mercury, so the title "baroscope" was given to these instruments with rough estimates of pressure. Lucien Vidie (1805–1866), a French physicist, is the inventor of the air barometer, officially called the "aneroid barometer." This barometer was the first of its kind to contain no liquid. It works by a metal cell containing only a very small amount of air, or a series of such cells joined together. Increased air pressure causes the sides of the cell or cells to come closer together. One side is fixed to the base of the instrument while the other is connected by means of a system of levers and pulleys to a rotating pointer that moves over a scale on the face of the instrument. As the chamber thickness changes due to pressure, it collapses and expands, moving the lever accordingly ("The Aneroid Barometer, and How to Use It", 1990). The aneroid barometer is significantly more compact and portable than the mercury barometer.

2.5 Anemometer

Flow of gases, air in motion, or wind continuously stirs in the atmosphere. Air moves vertically as well as horizontally; however, the speed of vertical displacements is usually a tenth or less of that of the horizontal component of wind (Anthes, 1997, p. 20). The vertical component of wind is primarily responsible for cloud formation and dissipation in the atmosphere. It is the horizontal component, however, that is regularly measured and recorded. Many instruments are used to measure wind velocity near the surface of the earth. The wind vane is mentioned as early as 48 BC, when it was recorded as being used in Athens, where it was believed that winds had divine powers (Denninger and Denninger, 2013). As the wind vane advanced throughout time it gathered many different shapes and sizes; however, its purpose remained the same, to indicate the direction that the wind is coming from. Once the basic concept of where the wind was coming from was known, the question of how fast or slow it was blowing became the next area of focus. An anemometer is a device that measures wind speed. The word "anemometer" is derived from the Greek word *anemos*, meaning wind. Current devices are much more advanced than the first invention; however, it is important to know how this instrument came about.

Sometime between AD 1200 and 1400, the Mayans created a pressure plate anemometer which consisted of small light balls that dropped from a basket and were blown downwind (Alcorn, 2002). In 1450, Italian architect Leon Battista Alberti (1404–1472) created the first mechanical anemometer. His anemometer was an elementary swinging plate in which a board swung along a graduated arc that was forced into the wind by a vane (Bud and Varner, 1998, p. 25). Several scientists, such as Robert Hooke and Santorio Santorio, created their own swinging plate anemometers and remarked on them in their works around the seventeenth century. Robert Hooke's work comments on the problem of resonance magnification. In 1805, Sir Francis Beaufort (1774–1857), an Irish hydrographer, designed a wind force scale, called the Beaufort Scale. He devised a scale using the observation of effects wind had on commonly used objects. The Beaufort Scale is used today mainly in regards to marine weather, and is also used to measure and describe the effects that different wind velocities have on various objects, whether on land or at sea. It should be noted that there are different terms to describe wind measurements, depending on the country in which one is in. In 1846, John Thomas Romney Robinson (1792–1882), an Irish astronomer and physicist, invented the four-cup hemispherical anemometer, which is still used today. The cups are designed to rotate horizontally with the wind, using a combination of wheels. As the device rotates, it calculates the amount of revolutions with respect to time. This is the premise of anemometers that we use today. During the twentieth century, three-cup anemometers were used and have since been modified to current technological standards.

Chapter Summary

Without the development of weather instruments, weather forecasting would not be accurate and feasible. These instruments aided in collecting data and statistical pattern developments. These devices gave concrete data, replacing divine interpretations and theories. And the innovative scientists behind the devices kick started a scientific revolution which began to supersede folklore and myths with facts, calculating, recording, and interpreting what was going on in our atmosphere. The combination of all these instruments helped open the doors to some extremely valuable new technologies, such as weather balloons, new mathematical approaches to forecasting, better communication, and devices for transmitting information over larger areas, all of which helped to create formalized weather departments and enhancements in communicating and publicizing weather forecasts.

References

Book References

Allaby, M. (1998) *Dangerous Weather: A Chronology of Weather*, Facts on File, Inc, New York.

Anthes, R.A. (1997) *Meteorology*, Prentice Hall, Upper Saddle River, NJ.

Bud, R. and Varner, D.J. (1998) *Instruments of Science: An Historical Encyclopedia*, Taylor & Francis, New York and London.

Reynolds, R., Hammond, J.C., Smith, F., and Tempest, S. (2008) *Eyewitness Companions: Weather-Forecasting, Weather Phenomena, Climate Change, Meteorology*, (eds. R. Gilbert, D. John, R.S. Colson, *et al.*), Dorling Kindersley Publishing, New York.

Journal/Report References

Biswas, A.K. (1967) The Automatic Rain-Gauge of Sir Christopher Wren, F.R.S., *Notes and Records of the Royal Society of London*, 22 (1/2), Royal Society, pp. 94–104.

Strangeways, I. (2004) Improving Precipitation Measurements. *International Journal of Climatology*, 24 (11): 1443–1460.

Strangeways, I. (2010) *A History of Raingauges*, TerraData Ltd., http://www.rmets.org/sites/default/files/pdf/presentation/20100417-strangeways.pdf, accessed 10 May 2015.

Website References

Alcorn (2002) *Hand-Held Anemometer*. Texas A&M University: Atmospheric Sciences, http://www.met.tamu.edu/courses/metr451/Ch6/Ch6-Part2/sld010.htm, accessed 11 May 2015.

"Barometer (Evangelista Torricelli)." (n.d.) *Evangelista Torricelli*. Purdue University, http://chemed.chem.purdue.edu/genchem/history/torricelli.html, accessed 11 May 2015.

Denninger, A.H. and Denninger, B.R. (2013) *A Brief History of Weather Vanes*. Denninger Weather Vanes & Finials, http://www.denninger.com/history.htm, accessed 11 May 2015.

Fowler, M. (2008) *Early Attempts to Understand the Nature of Heat*, University of Virginia, http://galileo.phys.virginia.edu/classes/152.mf1i.spring02/What%20is%20Heat.htmm, accessed 10 May 2015.

"In-Depth: Hygrometer." (2010) *Museo Galileo*, Institute and Museum of the History of Science, http://catalogue.museogalileo.it/indepth/Hygrometer.html, accessed 10 May 2015.

"The Aneroid Barometer, and How to Use It." (1990) *Bureau of Meteorology*. Commonwealth of Australia, http://www.bom.gov.au/info/aneroid/aneroid.shtml, accessed 11 May 2015.

The Editors of Encyclopaedia Britannica (2015) *John Frederic Daniell: British Chemist*. Encyclopaedia Britannica, www.britannica.com/EBchecked/topic/151016/John-Frederic-Daniell, accessed 10 May 2015.

Zimmerman, K.A. (2013) *Kelvin Temperature Scale: Facts and History*, Live Science, http://www.livescience.com/39994-kelvin.html, accessed 10 May 2015.

3

The Birth of Modern Meteorology (1800–1950)

Kevin Anthony Teague and Nicole Gallicchio

Once scientists had developed devices that provided them with concrete data, the new projected spectrum was communicating the data. Combining data in order to better understand earth's atmosphere began to generate new inventions, and simple weather forecasts were documented and publicized. Through new technological advancements more complex instruments and scientific methods were created and weather forecasting took a new turn in the nineteenth and twentieth centuries. The beginnings of modern weather forecasting began to take shape.

3.1   Telegraph

The means of communication before the nineteenth century was limited to the written word, word of mouth, and how fast your horse could travel. There were several different people who invented the workings of the electric telegraph around the same time, in various regions of the world, independently and not knowing of the others' work. In England, William Fothergill Cooke (1806–1879) and Charles Wheatstone (1802–1875) took out patents on their inventions in 1837 and 1845 respectively (Gribbin and Gribbin, 2003, p. 257). In America, long-distance communication was revolutionized during the 1830s by Samuel Morse (1791–1872), who had taken the invention to the fullest and made it into a practical form of communication.

Samuel Morse also invented the binary code known as Morse Code that enabled a form of unified communication. Created independently by several scientists and inventors, the telegraph was a device that worked by transmitting electrical signals through a wire connected to two stations.

Once the telegraph was invented, it enabled the meteorological community to expand its reaches of communicable weather forecasts. By the mid-1840s, meteorological observations and data were collected over widely scattered regions. The data and observations were shared via this new technology between Europe and North America. It was via these means of communication that meteorological and atmospheric observations were solidified. The means of sharing information between different time zones required weather to be recorded in Greenwich Mean Time. This allowed for data in different regions to be compared and processed. Realizing that most storms in North America formed and traveled west to east enabled scientists such as the Secretary of the Smithsonian Institution, Joseph Henry (1797–1878), in the mid-1840s, to propose a network of telegraphic links providing a warning system to citizens. However, thanks to Urbain Le Verrier's (1811–1877) initiative following the Crimean storm at the end of 1854, the French took the lead in Europe, setting up an official network of telegraphic weather warnings (Gribbin and Gribbin, 2003, p. 257).

The Evolution of Meteorology: A Look into the Past, Present, and Future of Weather Forecasting, First Edition.
Kevin Anthony Teague and Nicole Gallicchio.
© 2017 John Wiley & Sons Ltd. Published 2017 by John Wiley & Sons Ltd.

Not only did the telegraph change human communication, it altered the way we received information from weather balloons and kites (there is more on weather balloons in Section 3.4). The information was transmitted via a wire connected to the weather balloon which constantly emitted data. A Dutch scientist named E.H. von Baumhauer published a design in 1874 for a device called a telemeteorograph, which converted readings from a balloon or kite to a rotating drum (DuBois, Multhauf, and Ziegler, 2002, p. 16).

3.2 Fitzroy

Robert Fitzroy (1805–1865) was an English officer in the Royal Navy and a scientist. He is famous for being captain of HMS *Beagle* during Charles Darwin's voyage. Fitzroy was interested in weather for the purpose of saving lives, knowing first hand that storm warnings at sea saved his fellow mariners. He developed the fundamental techniques of weather forecasting and used that knowledge to create a prototype of a weather station. The station was much more basic than today's commonly seen weather stations, and consisted of a standard thermometer and barometer. Fitzroy used his training, weather stations, and observations to invent a system of storm warnings leading to his issuing of the first daily weather forecast. His work was published in the London *Times*, and he is said to be the inventor of the term "weather forecast" (Gribbin and Gribbin, 2003, p. 5). Later Fitzroy was "nominated to fellowship of the Royal Society for his hydrographic and chronographic survey, and was also chosen as the first Chief Statist of the newly formed Meteorological Department of the Board of Trade in the UK" (Sibley, 2005).

3.3 Hugo Hildebrand Hildebrandsson

Hugo Hildebrand Hildebrandsson (1838–1925) was a Swedish meteorologist and first professor of meteorology at Uppsala University. He was also part of the Clouds Commission of the International Meteorological Committee. His research was physics based and specific to the subject of clouds. His many papers on the centers of action of the atmosphere mark a great advancement in seasonal forecasts as well ("Hugo Hildebrand Hildebrandsson," 2011). In 1896, Hildebrandsson wrote a book called *International Cloud Atlas*, to support the training of meteorologists in order to promote better weather forecasting and a greater understanding of clouds. This book contains photographs and detailed descriptions of clouds, and even though it has been edited and republished numerous times since it was first written, the *International Cloud Atlas* still aids in the education off meteorology today.

3.4 Weather Balloons/ Radiosondes

Prior to the invention of the weather balloon in the late eighteenth century, rockets and kites were the only two vehicles capable of ascending into the atmosphere. Rockets have been used in China and Europe for military purposes since the early thirteenth century, while kites were known simply as a child's toy, but later also became a valuable tool for meteorology. The first attempt at an atmospheric probe began in 1749 by a British astronomer, Alexander Wilson, when he attached thermometers onto kites in order to measure temperatures at higher altitudes. Wanting to gather information at higher altitudes, the use of balloons was recognized by the late 1780s. The main issue with a balloon was that its movement was problematic. This flaw was rectified in the 1890s with the Siegfried and Parseval box kites, although these could not reach as high as the balloon (Figure 3.1). Using the late-eighteenth-century concept of the French physicist Jacques Charles (1746–1823), meteorologists began to substitute the hot air inside the balloons with hydrogen (DuBois, Multhauf, and Ziegler, 2002, p. 3).

Figure 3.1 Photo estimated to be from 1894, showing weather kites in action. Weather kites were used to gather weather data, including temperature, humidity, pressure, and winds, from higher levels. This was not an accurate or easily controlled form of data collection, leading to the expansive use of weather balloons. *Source*: Photo courtesy of NOAA Photo Library.

The success of this style of balloon led to the development of balloonsondes.

Early balloonsondes were non-telemetering, unmanned, coal-gas or hydrogen-filled balloons, able to carry self-registering thermometers and barometers. Radiosondes were developed following the success of the telegraph, allowing meteorologist to relay information from the balloon above to a ground station. A radiosonde is a combination of weather instruments that measure the atmospheric conditions aloft by a balloon and transmits its data and readings by radio to ground level. They are called radiosondes because they take "soundings" and transmit them to receiving stations via radio (Allaby, 1998, p. 10). The balloons travel approximately 24,000 m (78740 ft) into the middle of earth's stratosphere. Most balloons are approximately 1.5 m (5 ft) in diameter and filled with hydrogen. While the balloon is in flight, the radiosonde transmits measurements back to the ground station. As the balloon rises 4.5 m/s (14.7 ft/s) the hydrogen expands and bursts at a height of approximately 24,000 meters. The weather instruments then parachute back down to the ground, and during the nineteenth century they were sometimes not found for days due to the inability to track the descent of the parachute. Today, radiosondes have tracking devices that allow the returning parachute containing the instruments to be found easily when over land, otherwise GPS tracking maps the exact coordinates and path that the balloon takes during its flight. Radiosondes have deepened our understanding of upper-atmospheric trends and have made significant impacts on weather forecasting.

3.4.1 Barothermograph/ Barothermohygrograph

Sending individual meteorological instruments into the upper atmosphere became costly and inefficient. In order to collect all meteorological data, scientists had to incorporate all the instruments into one multipurpose device. In 1898, the Richard brothers of France developed a barothermograph. This instrument combined a thermometer and a barometer and continuously collected and recorded data while in flight in the upper atmosphere (DuBois, Multhauf, and Ziegler, 2002, p. 4). A barothermohygrograph incorporated the barometer, thermometer, and hygrometer, which was proven to be most efficient since it contained arguably the three major and most important forecasting tools.

3.5 Birth of Governmental Weather Departments

Early methods of forecasting were never sound and/or accurate enough to be considered as part of everyday needs. This was partly due to a lack of knowledge on the subject of meteorology and atmospheric sciences but also due to the lack of technology that allowed for growth of knowledge in those fields. After the meteorologically based inventions previously discussed, such as the thermometer and barometer, the ability to put together a better understanding of weather was finally established. With this new ability to make educated and scientific predictions and evaluations of the weather conditions, as well as the new ability to communicate this information, with the invention of the telegraph, a more modern style of weather forecasting was eventually born. This new style of weather gathering and forecasting became more pronounced and more valuable in day-to-day life. Everything that occurred prior to 1854 led to the birth of the first formalized National Weather Service, the Meteorological Department of the Board of Trade of the United Kingdom, better known today as the Met Office (Walker, 2012).

The Met Office began as an office of four staff members, and Robert Fitzroy, mentioned above, was chosen to head the department. Fitzroy and his staff were under a lot of scrutiny from the beginning. Data gathering and analyzing were the main objectives early on, with many days' worth of data observations and logbooks on weather conditions created and analyzed. By 1860, collections of weather reports were distributed by telegraph from 15 coastal stations in and around the United Kingdom (Figure 3.2), These weather reports were created from observations made each day at 8 a.m. and transmitted to the Met Office in London. These reports included weather conditions such as temperature, wet bulb temperature, barometer readings, wind direction, and wind speed. By 1861, the Met Office began publishing weather forecasts for the public (Monmonier, 1999, pp. 44–46). During this time, the Met Office created forecasts that predated the numerical process that was developed by Richardson (see Section 3.6), his followers, and computers. Instead, forecasts were mainly based on past weather experiences with similar characteristics to the weather at the present, leading to vast inaccuracies. By 1866, these inaccuracies forced a 13-year prohibition on the Met Office's forecasting for the public (Walker, 2012).

By 1870, the United States began putting more value on weather and the science that encompasses it and President Ulysses S. Grant authorized the Secretary of War to establish a national weather service, the Weather Bureau, better known today as the National Weather Service ("History of the National Weather Service," 2012). The United States and the United Kingdom were not the only countries developing their own nationalized and governmentally run weather departments. The development of these departments began to pop up across the world, ranging from but not limited to Japan, Denmark, the Netherlands, Argentina, Australia, Canada, and then even later on into the 1900s, Poland, France, and Germany. Although many nations were developing their own weather services, the main contributors to

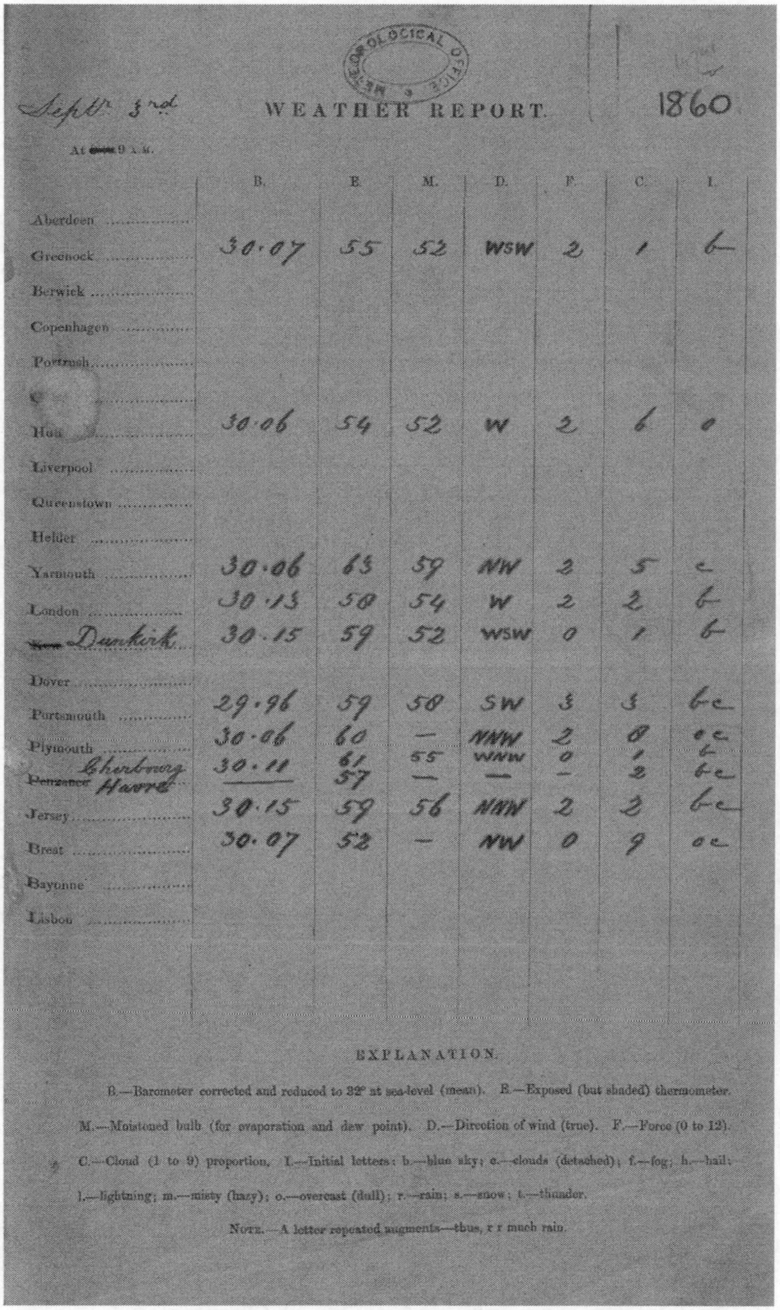

METEOROLOGICAL OFFICE

Septr 3rd WEATHER REPORT. 1860

At 9 A.M.

	B.	E.	M.	D.	F.	C.	I.
Aberdeen							
Greenock	30·07	55	52	WSW	2	1	b
Berwick							
Copenhagen							
Portrush							
[illegible]							
Hull	30·06	54	52	W	2	6	o
Liverpool							
Queenstown							
Heldier							
Yarmouth	30·06	63	59	NW	2	5	c
London	30·13	58	54	W	2	2	b
~~Nes~~ Dunkirk	30·15	59	52	WSW	0	1	b
Dover							
Portsmouth	29·96	59	58	SW	3	3	bc
Plymouth	30·06	60	—	NNW	2	8	oc
Cherbourg	30·11	61	55	WNW	0	1	b
~~Penzance~~ Havre	—	57	—	—	—	2	bc
Jersey	30·15	59	56	NNW	2	2	bc
Brest	30·07	52	—	NW	0	9	oc
Bayonne							
Lisbon							

EXPLANATION.

B.—Barometer corrected and reduced to 32° at sea-level (mean). E.—Exposed (but shaded) thermometer. M.—Moistened bulb (for evaporation and dew point). D.—Direction of wind (true). F.—Force (0 to 12). C.—Cloud (1 to 9) proportion. I.—Initial letters: b.—blue sky; c.—clouds (detached); f.—fog; h.—hail; l.—lightning; m.—misty (hazy); o.—overcast (dull); r.—rain; s.—snow; t.—thunder.

Note.—A letter repeated augments—thus, r r much rain.

Figure 3.2 This is likely the very first weather report from Vice-Admiral Robert Fitzroy of the Meteorological Office (today known just as the Met Office). The report was from 3 September 1860. This report includes location, air pressure, temperature, wind direction, wind speed/force, cloud coverage, and type of weather occurring. *Source*: http://www.metoffice.gov.uk/about-us/legal/tandc#Use-of-Crown-Copyright. Reproduced with permission of Crown-Copyright.

the early advancements of the nationalized weather system were the Met Office of the United Kingdom and the Weather Bureau of the United States. Due to this boom of weather departments, in 1873 the International Meteorological Organization was created, combining the knowledge, research, and advancements from many nations all working together to further the field of meteorology. As of 2013, this organization is a United Nations Agency and is known as the World Meteorological Organization (WMO), with 191 member states and territories ("WMO in Brief," n.d.).

This growth in the field of meteorology was greatly seen by the rapid developments within the Weather Bureau of the United States. By the late 1870s, the United States had 22 telegraph stations connected to Washington DC, providing weather updates and observations (Figure 3.3).

Eventually, this type of information made it into the first regular newspaper weather map in the country in the *New York Daily Graphic* in 1879 (Monmonier, 1999, pp. 157–159). In 1890, the Weather Bureau became a civilian agency, and in 1898 President William McKinley established the Hurricane Warning Network in the West Indies. This network eventually became part of the system that forecast the great hurricane of 1900 which devastated Galveston, Texas. ("History of the National Weather Service," n.d.). It wasn't until 1900 when the Weather Bureau officially moved from the War Department to the Agricultural Department of the government (Barnett, 2015, Section II) (Figure 3.4).

A huge leap in accuracy and services occurred in 1909 when the Met Office created a way for ships to wirelessly telegraph observations and forecasts. This led to quicker transmissions of

Figure 3.3 Photo of an 1880s weather station, located on Pikes Peak, Colorado. *Source*: Courtesy of NOAA Photo Library.

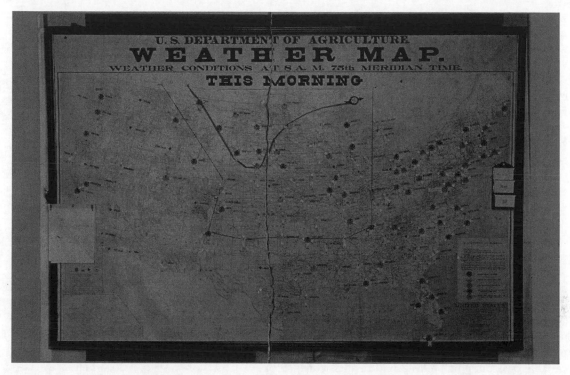

Figure 3.4 Photo of a 1900 daily weather map from the Weather Bureau, today known as the National Weather Service. *Source*: Courtesy of NOAA Photo Library.

weather updates due to near real-time observations ("Overview of the Met Office," 2012). This is also around the time the Weather Bureau began to use weather balloons to gather data on the upper atmosphere, as stated in Section 3.4. By 1918, the Weather Bureau began sending weather bulletins for domestic military flights as well as for airmail services ("History of the National Weather Service," n.d.).

3.6 Numerical Weather Prediction

Another attempt to increase the accuracy of forecasting occurred when the theory of thermodynamics was developed toward the end of the nineteenth century. Thermodynamics was one of the final pieces in understanding the physics of the atmosphere. By 1890, an American meteorologist named Cleveland Abbe concluded

that "meteorology is essentially the application of hydrodynamics and thermodynamics to the atmosphere" (Lynch, 2007, p. 2). It was at this point that mathematics went from a background contributor to weather forecasting and moved into the driver's seat. This style of thinking became stronger when a Norwegian scientist, Vilhelm Bjerknes, created a strategy for "rational forecasting." This strategy included a "diagnostic step," in which the initial state of the atmosphere was needed to be collected, and a "prognostic step," in which the laws of motion were calculated to show the changes that were to occur over time. The final contributor to this new approach and forecasting method was an English scientist employed by the Met Office, Lewis Fry Richardson. Richardson, using the concept of Bjerknes with some of his own theories and reasoning, put his method to the test in 1922. He developed and published his results of a forecast created by using the most available weather observations as his initial data and a numerical method that he created to calculate the changes in pressure and winds. His results were terribly inaccurate and physically impossible. Richardson blamed the errors on incorrect initial data, but was still heavily criticized for his inaccuracies.

Although the results were inaccurate, the vision happened to be precise. The method developed by Richardson turned out to be too large and complex for its time, making it impossible to tackle and took far too long to get to any results. Often times, weather forecasts were completed hours and even days after the weather event had taken place. Richardson stated that in order to perfect his approach he would need 64,000 people to keep up with the pace of the atmosphere (Lynch, 2007). It wasn't until the invention of computers in the mid-twentieth century that Richardson's dream of 64,000 people came true; those 64,000 people being the processors inside the supercomputers that were capable of producing split-second results while taking in tons of data and information. This method is what is now used on a daily basis and is discussed further in Chapter 4 (Section 4.4).

3.7 Weather Broadcasting

The year 1919 saw the foundation of the American Meteorological Society (AMS), which was created to supplement the information dispersed by the Weather Bureau. The AMS eventually played a large role in the further development of the field of meteorology, with its members consisting mostly of professionals educated in the field of meteorology and certified for consulting and broadcasting of meteorology (see Chapter 10, Section 10.1) ("History of the AMS," 2015). By 1921, the first weather broadcast done by radio in the United States was relayed from the University of Wisconsin, and was later credited for being the beginning of the "weatherman" (Barnett, 2015, Section II). The United Kingdom followed suit with its first public weather radio broadcast in 1922 by the

BBC (the British Broadcasting Corporation), and by 1923, daily weather forecasts were broadcast by radio and have been ever since. This increase in means of relaying forecasts continued to the marine sector when by 1925 the Met Office was sending information to be broadcast to ships in the Maritime and Coastguard Agency ("150 Years of Weather Forecasting for the Nation," n.d.). The United States continued advancing as well, and by 1926 the Weather Bureau formed civilian aviation weather services. By the late 1920s, land, water/marine, and aviation forecasts were being created and broadcast. The Weather Bureau continued to grow and developed its own Hurricane Warning System in 1935, and by 1940 the government transferred the Bureau to the Department of Commerce as the effects weather had on trade became more evident ("History of the National Weather Service," n.d.).

The BBC then moved into televised broadcasting and created the very first TV broadcast of a weather forecast in 1936, when it televised the first weather map, consisting of a hand-drawn weather chart with isobars and temperatures displayed. In 1941, the United States had its first ever televised weather forecast on what would later be the NBC network in New York City. It consisted of a screen of text and no map or charts (Monmonier, 1999, pp. 177–179).

3.8 Forecasting for WWII

By the 1940s, weather forecasting became an important part of people's everyday life, whether by newspaper, radio, or television. The value of weather forecasting was not only for the public but also for military purposes, as seen in World War II. With the Met Office already part of the Ministry of Defense due to its importance in World War I for giving the military an edge with its weather forecasts, it played one of the largest and most unsung roles in the D-Day Landings of WWII in 1944. General Eisenhower, the commander of the Allied forces, changed the date for the Allied invasion of France in 1944 following advice from the RAF meteorologist Captain James M. Stagg, who pinpointed when there would be a break in the weather of sufficient duration to allow the invasion force to get across the English Channel. This forecast was extremely accurate, allowing the Allied forces to make the journey safely and land at a time German troops were not expecting because of the weather conditions they were experiencing at the time. Captain Stagg's success eventually earned him the position of Director of the Met Office (Walker, 2012).

3.9 Extreme Weather Forecasting: 'Nowcasting'

As the pre-computer history of weather forecasting and meteorology drew closer to an end in the late 1940s, the United States experienced possibly the beginning of what is now known as "nowcasting" – forecasting on current, up-to-the-minute weather conditions as they occur – and "extreme weather forecasting." In 1948, the first ever tornado warning was issued, for Tinker Air Force Base in Oklahoma. Five days prior, a series of storms moved through the same area dropping a devastating tornado in a nearby town. Surprisingly, five days after this storm system, weather conditions seemed to be a carbon copy of those same tornadic storms. Weather conditions and observations were scrutinized in intense detail and, by morning, weather charts continued to show great similarities to the prior storms. As clouds and thunderstorms began to develop during the afternoon, meteorologists mapped and estimated the storms' location and timing for when they would move to the Tinker Air Force Base area. After hours of nail-biting

Figure 3.5 The aftermath of the 25 March 1948 Tinker Air Force Base tornado. The days and hours leading up to this tornado saw the first ever tornado warning, saving lives from the destruction of this tornado. *Source*: Courtesy of NOAA Photo Library.

analysis, meetings, and discussions, Captain Robert Miller and Major Ernest Fawbush decided to type up and distribute the warning to the Base Operations for it to be sent out. The tornado did occur, destroying the base, but causing zero fatalities. The first ever tornado warning was a huge success and is credited for saving many lives in the path of that storm system ("The Historic Forecast," n.d.) (Figure 3.5).

Section Summary

By 1950, weather forecasting, meteorology, and atmospheric science were no longer based on folklore, guesswork, or works of the gods. Thanks to their development, the weather could be calculated, explained, and even predicted. Meteorology took a long time to develop, and thousands of years for governments to realize its significance, but when technology and industrialization helped with the advancement of the technology behind weather instruments and the forms of communicating its data, and when new theories and mathematical processes were developed to increase the accuracies of forecasting, meteorology became an extremely important field of science to the world. So much importance was placed on weather forecasting and weather knowledge that more and more advancements were strived for to improve all areas of the science. These advancements would help shape our technical understanding of meteorology today.

References

Book References

Allaby, M. (1998) *Dangerous Weather: A Chronology of Weather*, Facts on File, Inc, New York.

Barnett, C. (2015) *Rain: A Natural and Cultural History*, Crown Publishers, New York.

DuBois, J., Multhauf, R., and Ziegler, C. (2002) *The Invention and Development of the Radiosonde, with a Catalog of Upper-Atmospheric Telemetering Probes in the National Museum of American History: Smithsonian Institution*, Smithsonian Institution Press, Washington DC.

Gribbin, M. and Gribbin, J. (2003) *FitzRoy*, Headline Book Publishing, London.

Monmonier, M. (1999) *Air Apparent: How Meteorologists Learned to Map, Predict, and Dramatize Weather*, University of Chicago Press, Chicago.

Walker, M. (2012) *History of the Meteorological Office*, Cambridge University Press, New York.

Journal/Report References

Lynch, P. (2007) The origins of computer weather prediction and climate modeling. *Journal of Computational Physics*, 227 (7), pp. 3431–3444.

Website References

"150 Years of Weather Forecasting for the Nation." (n.d.) *Met Office*, Met Office, http://www.metoffice.gov.uk/media/flash/s/s/MetOffice.swf, accessed 1 May 2015.

"History of the AMS." (2015) *American Meteorological Society*, AMS, http://www2.ametsoc.org/ams/index.cfm/about-ams/ams-history/history-of-the-ams/, accessed 1 May 2015.

"History of the National Weather Service." (n.d.) *National Weather Service*, NWS & NOAA, http://www.weather.gov/timeline, accessed 1 May 2015.

"History of the National Weather Service." (2012) *Public Affairs Office*, NWS & NOAA, http://www.nws.noaa.gov/pa/history/, accessed 1 May 2015.

"Hugo Hildebrand Hildebrandsson." (2011) *Encyclopedia Britannica 1911*, ITA, www.

theodora.com/encyclopedia/h2/hugo_
hildebrand_hildebrandsson.html, accessed
27 April 2015.

"Overview of the Met Office." (2012) *Met Office*,
Met Office, http://www.metoffice.gov.uk/
news/in-depth/overview, accessed 1 May 2015.

Sibley, A. (2005) *FitzRoy: Captain of the Beagle,
Fierce Critic of Darwinism*, Institute for
Creation Research, http://www.icr.org/
article/fitzroy-captain-beagle-fierce-
critic-darwinism, accessed
11 May 2015.

"The Historic Forecast." (n.d.) *Tornado
Forecasting 50 Years*, NOAA, http://www.
outlook.noaa.gov/tornadoes/torn50.htm,
accessed 1 May 2015.

"WMO in Brief." (n.d.) *About Us: World
Meteorological Organization*, WMO, http://
www.wmo.int/pages/about/index_en.html,
accessed 1 May 2015.

Figure References – In Order of Appearance

"NOAA Photo Library." (n.d.) *NOAA*, http://www.
photolib.noaa.gov/htmls/wea01100.htm,
accessed 18 August 2016.

"Robert FitzRoy and the Daily Weather Reports."
(2016) *Daily Weather reports*. Met Office,
http://www.metoffice.gov.uk/learning/library/
archive-hidden-treasures/daily-weather-
reports, accessed 10 July 2016.

"NOAA Photo Library." (n.d.) *NOAA*, http://www.
photolib.noaa.gov/htmls/wea00908.htm,
accessed 18 August 2016.

"NOAA Photo Library." (n.d.) *NOAA*, http://www.
photolib.noaa.gov/htmls/wea01801.htm,
accessed 18 August 2016.

"NOAA Photo Library." (n.d.) *NOAA*, http://www.
photolib.noaa.gov/htmls/wea00224.htm,
accessed 18 August 2016.

Section II

The Early Television, Computer, Satellite, and Radar Era (1950–1980)

4

Television and the First Computerized Advancements

Kevin Anthony Teague

By the 1950s, meteorology shifted toward becoming a technology- and computer-driven science. The 30 years following World War II no longer consisted solely of weather forecasts strictly based on knowledge of prior weather events, dispersed primarily by radio or telegraph. Instead, the mid-twentieth century consisted of household televisions which took over the distribution of forecasting to the public, and supercomputers that could calculate astronomically sized formulas and data in seconds to create more accurate forecasts. Adding to the advancements during this time were new tools for meteorologists in the form of radar and satellite. Once the structure of the field of meteorology became established in the first half of the twentieth century, it blossomed, from an exercise in folklore as it had been in ancient times to a science. Forecasts became more accurate and geographically specific, and the science of forecasting spread rapidly as more government agencies continued to develop it. This section demonstrates this rapid growth and the impacts it had on future developments within the field of meteorology.

4.1 Television in the Household

By 1950, television was starting to become commercialized and slowly became more readily available in the general household. In 1950, 9%

of American households had a television. By 1960, this percentage jumped dramatically to 90% ("Moving Image Section: Motion Picture, Broadcasting and Recorded Sound Division," n.d.). Worldwide there was a similar jump. In 1953, 89% of all television receivers in the world were found in North America, most being in the United States. Europe accounted for nearly the rest of the remaining 11%. By 1960, the dramatic jump in television sets also occurred outside the United States, with North America accounting for only 60% (even though Americans themselves were previously at nearly 90%, as stated earlier), Europe jumped up to 31%, and the rest of the world started to become part of the television phenomenon, accounting for the remaining 9% (UNESCO, 1963, p. 23). With this jump in the number of televisions came an opportunity for weather forecasts to be broadcast across larger areas to inform the public on a much greater scale and much more regularly.

4.2 Television Weather Forecasting of the 1950s

With television becoming a household staple, broadcasting weather forecasts on air became a more regular practice as well. After World War II, many war veterans tried their luck at weather forecasting on television. One of those war veterans, Clint Youle, made the first national

The Evolution of Meteorology: A Look into the Past, Present, and Future of Weather Forecasting, First Edition.
Kevin Anthony Teague and Nicole Gallicchio.
© 2017 John Wiley & Sons Ltd. Published 2017 by John Wiley & Sons Ltd.

appearance on NBC in Chicago in May of 1949, reporting the weather one to three times a week. Youle reported the weather using a map purchased at a local store with a piece of Plexiglas in front of it, allowing him to be able to draw and erase weather symbols and words (Ravo, 1999). After Youle, TV weather forecasting took off and never looked back. Originally, weather forecasting on the news was mainly geared to those whose livelihoods were affected by the weather, such as farmers and travelers. These forecasts, as well as those of national networks, were usually reported by war veterans, like Youle, or by retired Weather Bureau employees. Other more localized stations took a different approach. In the United States, the 1950s were the time of the "weather girls." These on-air weather reporters used their time reporting the weather in a comedic act or show-like manner. The "weather girls" were carefully selected for their smile, personality, and looks. The first of them was Carol Reed, who reported for 12 years on WCBS New York (Henson, 2010, p. 110). It was commonplace for weather reporting to be seen as a joke. Because of this, the American Meteorology Society decided to establish its Seal of Approval Program in 1957. This program was designed to "promote understanding and reliability" within weather forecasting and reporting (Monmonier, 1999, p. 179).

In Europe, weather reporting first occurred in January of 1954, when George Cowling presented the first live weather forecast on the BBC. Similarly to Youle, Cowling stood in front of a map and drew his forecast and symbols directly onto it. Cowling stayed reporting for the BBC until 1957, but remained with the Met Office well into the 1980s. And just like Cowling, all of the BBC's meteorologists who followed him were also from the Met Office ("150 Years of Weather Forecasting for the Nation," n.d.).

By the mid- to late 1950s, a more professional approach to weather reporting took hold. During this time, the standard format for TV weather forecast and reports was created and has remained relatively similar to this day. This format starts with the location's current conditions. Next are the previous high and low temperatures. Following these is the current map with the forecast map after that. Finally, the segment ends with the local forecast (Monmonier, 1999, p. 183). This format still holds true with more modern alterations due to growth in technology and increased visual aids, such as radar and satellite imagery, and weather models. These technology aids began to be seen first in the United States. As the Weather Bureau was trying to develop and deploy weather radars, a couple news stations beat them to it and added their own radar to their weather-reporting repertoire. The first two stations that used radar were located in New Orleans (in 1954) and Cincinnati (in 1956). Shortly after that, radar became a standard tool in weather reporting, especially in areas that suffered frequent severe weather events such as in the Midwest and the Great Plains (Henson, 2010, p. 89).

4.3 Television Weather Forecasting of the 1960s and 1970s

The 1960s and 1970s reflected an even greater development of weather technologies and science across the world. In 1963, European TV and the Met Office itself began developing and using radar in their weather reports. At nearly the same time, the United States began to supplement its radar resources with yet another tool, using the first ever satellite imagery inside weather forecasts. This satellite imagery was used first by two US stations in Florida, as they pieced together satellite images from TIROS I (a US satellite launched in 1960). Reports took still images from the satellite and, in cartoon-like fashion, put images together to try to format a loop of cloud movement (Henson, 2010, p. 92). By 1970, European television weather reports began to utilize satellite imagery technology in their own on-air reports.

By 1975, the Geostationary Operational Environmental Satellite (GOES), the first geostationary satellite, was launched and in action allowing weather reports to loop higher-quality images of the clouds and weather system movements (see Chapter 5 for more detailed information on radar and satellites).

Most TV forecasts were black and white through the 1960s until around 1970, when color televisions became increasingly more common. In the United States alone, 39% of televisions in homes were color in 1970. By 1978, this number jumped to 78% (Henson, 2010, p. 91). Along with color, the weather graphics displayed also began to change. Up through the early to mid-1970s, weather graphics consisted mostly of dull information from governmental charts, crude maps, hand-drawn weather symbols, and then eventually symbols that were magnetic that could be moved around the map by hand. Showing the weather instead of just reporting on the weather was a much more stimulating and understandable approach. The sudden boom in radar and satellite technology, along with color televisions, improved the quality of weather reporting, with increasingly visually stimulating pictures and images enticing viewers. Weather forecasting became a highly desired and anticipated portion of a television broadcast. Weather reporters became recognized as part of the news family and news team. BBC made European history in 1974 when they employed their first ever woman weather reporter, Barbara Edwards. This was of huge significance socially throughout Europe, as Edwards' employment marked a shift in this heretofore male-dominated occupation.

By 1978, computer costs were still too pricey for news stations to utilize; however, there were two stations in the United States (Tampa and Madison) that came together and created WeatherVision, the nation's first commercial network affiliation. This weather graphic system was the pioneer in TV weather graphics and started the trend that was further developed in the 1980s (Henson, 2010, p. 96).

4.4 The Beginning of Computers and Numerical Weather Prediction (NWP)

As television grew from the 1950s through to the 1970s, on-air meteorologists started to rely on forecasts created by computer models. These computer models were developed after the work of Richardson in the 1920s (see Chapter 3, Section 3.6). When Richardson developed his numerical weather prediction formula, he was unable to calculate all his data fast enough or accurate enough to predict weather events, let alone finish his forecast before the storm took place. As stated in Chapter 3, Richardson said that he would need 64,000 people to calculate and formulate a forecast using his process. It wasn't until the 1930s and 1940s, when the top mathematician in America, John von Neumann, got involved, that Richardson's dream was finally able to be put to the test. John von Neumann was involved with many different mathematical topics, including areas such as logic, abstract algebra, and quantum physics. When von Neumann began to study about turbulent fluid flows, he realized that a machine would be needed to calculate his massive equations. Von Neumann masterminded the construction of an electric computer around 1950, in what he called the Electronic Computer Project (Lynch, 2007).

Prior to his involvement with computers, von Neumann previously followed Richardson's work on NWP (numerical weather prediction). It was his knowledge of NWP that made him decide that meteorology and weather forecasting were prime candidates to test his computer on. In 1950, meteorologist Jule Charney and his team ran their meteorological algorithms through the only available computer at the time, the Electronic Numerical Integrator and Computer (ENIAC) (Figure 4.1) and produced the groundbreaking first ever computerized forecast. This forecast consisted of a 24-hour forecast, which ended up taking 24 hours to complete. While this was a huge achievement,

Figure 4.1 Photo of ENIAC with the core memory stored in the white cabinet. *Source*: Photo courtesy of US Army Photo from the archives of the ARL Technical Library.

the forecasts that were produced still only kept pace with the weather, and did not go ahead as true forecasts should. Regardless of the time it took to create the forecasts, this positive result of a computerized NWP forecast created a worldwide interest in computer-based forecasting (Lynch, 2007).

By 1966, Charney and many others were focusing on a main issue that was causing inaccurate forecasts, as was the case with Richardson himself in the 1920s: the issue of data initialization. Charney was able to have success producing a good simulation of "development, occlusion, and frontal structures," and was able to produce a six-level primitive equation model

which he ran on a CDC6600, the first successful supercomputer, yielding very positive results. In 1972, this model was increased to a 10-level model and incorporated for the first time a "sophisticated parameterization of physical processes," leading to the first ever useful precipitation forecast (Lynch, 2007).

Forecasting skill scores at the birth of the computer in the early 1950s were around 35% and by the 1960s with the one-layer model, skill level rose to near 50%. In the mid-1960s the three-layer model produced skill scores of nearly 55%, only to be surpassed by the six-layer model around 1970, with scores of over 60%. It wasn't until the higher-resolution computers and the eventual 12-layer model in the late 1970s and early 1980s that skill scores of over 80% occurred. From the birth of computers, through the 1970s, skill scores grew from 35% to nearly 80%, and this growth coincided with the power growth of the supercomputers (Monmonier, 1999, p. 96).

4.5 Computers and NWP in the Met Office

Shortly after von Neumann's success in America, the Met Office began using an electrical desk calculator to work on Richardson's original formulas. By the mid-1950s, the Met Office acquired Leo, the world's first business computer machine with enough power to process the data required for a weather forecast ("Leo Computers Society," 2015). By 1959, the Met Office bought its next computer, called a Ferranti Mercury, or Meteor. The Meteor was the answer to Richardson's dream, with the computer having the capability of performing 30,000 calculations per second, one of the largest and fastest computers in all of England at the time ("Supercomputers," 2014). As a better understanding of the atmosphere formed, new ways of calculating the ever-changing atmospheric dynamics became necessary, resulting in new computer codes and data which required newer and more powerful supercomputers. By 1965, the Met Office bought an English Electric KDF9, which had upped calculation speeds to 50,000 calculations per second ("Supercomputers," 2014) (Figure 4.2). It was with computers like this that the Met Office and areas such as Heathrow Airport were able to do away with manual forecasts for upper air and flight mapping, and replace them with the computerized forecasts that ran 24 hours a day seven days a week (Golding, Mylne, and Clark, 2004, p. 300).

A lot of training and specialized knowledge was required to understand and run these supercomputers. Along with that, computer power constantly ended up running short of the needs of modelers and forecasters. As new information about weather and the atmosphere was realized, and new mathematical approaches learned and discovered, newer codes and algorithms had to be produced and run on more powerful computers. This pattern continued usually every three to five years (Edwards, 2010, pp. 171–172). By the early 1980s, the Met Office's supercomputer calculation speed had increased from 30,000 calculations per second in the 1950s to 200 million calculations per second. To put this into perspective, today's computers are capable of producing over 10,000 trillion calculations per second ("Supercomputers," 2014).

4.6 Computers and NWP Worldwide

The computer explosion was not limited to just the Weather Bureau and the Met Office. Worldwide contributions and achievements were made, starting in the 1950s, all stemming from the usage and computerized implementations of the NWP. Computers were developed and used in nations such as Italy, Canada, Algeria, New Zealand, Australia, China, and many more. Three nations of focus were Japan, Germany, and France.

In the late 1940s and into the 1950s, the Japan Meteorological Agency (JMA) performed a lot of research and work to create and expand the NWP. By the late 1950s, with very little funding from the federal government, JMA meteorologists did most of their calculations on graphs, desk calculators, and an electro-mechanical relay-switching computer called FACOM. By 1959, JMA obtained the IBM704, one of the first mass-produced computers in the world, and this acquisition resulted in a formalized NWP branch of its agency. The IBM704 allowed the JMA to create forecasts running to 48 hours. Although this forecast had many flaws, it was the foundation of great things to come for Asian weather services (Persson, 2005, pp. 270–271).

German weather services began running forecasts up to 72 hours on a Swedish-made

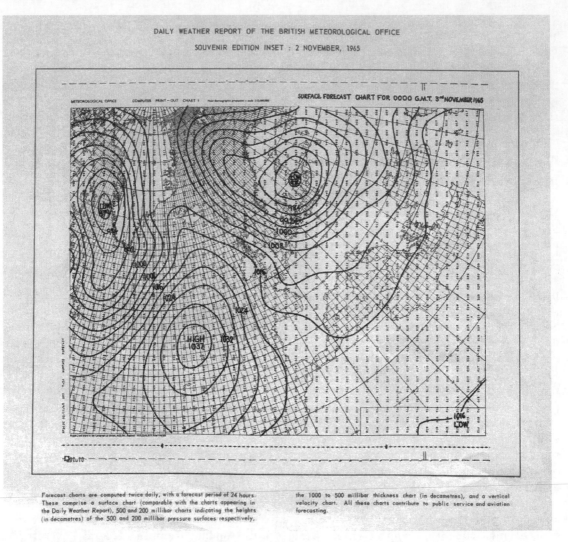

Figure 4.2 On 2 November 1965, the Met Office issued its first NWP computer weather report. This began the surge of computerized modeling through to the present day. Notice the lack of detail and layers in the mapping on this state-of-the-art computer model, nothing close to the imagery produced by the Met Office today. *Source*: Image provided by the kind permission of the Met Office.

computer called BESK, in 1955. These forecasts were baroclinic forecasts, which means they worked on the idea of the atmosphere having different air mass regions, with cold fronts and warm fronts ushering in different air masses as well as the idea of jet streams, troughs, and ridges. The baroclinic forecasts developed by the Germans at this time forecast at the 850, 700, 500, and 300 hPa levels. In order to better calculate and create their forecasts, the Germans also switched to the IBM704 and, by 1966, an operational NWP 72-hour barotropic forecast (meaning a model that consisted of a uniform temperature distribution, with no fronts or jet stream air mass differences) was created with three dimensions and a resolution of 381 km, which was extremely impressive for the time. In 1969, a Russian academician, E. K. Fedorov, strongly supported the sharing of technology with the former German Democratic Republic. The Soviets ended up giving them the BESM-6, a computer which needed its own building to store and operate, as well as an extensive training to use it. By the mid-1970s, routine NWP was run with a model consisting of four levels: 850, 700, 500, and 300 hPa, and an even stronger resolution of 300 km. This model was run twice a day and was extremely detailed and elaborate for the time, providing "extreme temperature, precipitation, the duration of sunshine, and surface winds up to 45 hours" (Persson, 2005, pp. 272–273).

Météo-France, the national weather service of France, began to use computers in 1954, with the French machine CAB2022. This computer allowed the French to create a simple barotropic model of 24 hours, 48 hours, and 72 hours, but the poor technology of this computer created many bugs and breakdowns. In 1960, Météo-France began using the KL901 computer. This computer was brought in as a prototype, and was used for "barotropic modeling, automatic plotting, chart drawing, and analysis" (Persson, 2005, p. 275). In 1965, two Frenchmen, Jean Lepas and Jean Labrousse, were sent to the United States to study the Americans' Weather Bureau and their implementation of NWP, and to gather information on computer manufacturers. This was a time when numerous meteorologists and scientists were sent abroad to gather information and to learn from other nations. By 1966, new computers were being used by France, including the CDC6400, and they were able to create their first primitive equation model by the mid-1970s, in which they ran a daily barotropic, 10-level model, which was made available to forecasters in Paris.

4.7 The 1970s and Its Lasting Influences on Forecasting

The 1970s had a few more major advancements when it comes to forecasting and computer technology. Possibly the greatest accomplishment was the creation of MOS, or model output

statistics, in 1972 by Harry R. Glahn and Dale A. Lowry. MOS "consists of determining a statistical relationship between the predicted and variables from the numerical model at some projection time" and is the "determination of 'weather-related' statistics of a numerical model" (Glahn and Lowry, 1972, p. 1203). MOS has been developed and used widely (and currently, by the National Weather Service, formerly the Weather Bureau up until 1970) to prepare guidance forecasts across the whole United States. MOS data are collected and calculated at and for a fixed point or station. There are many stations across the United States, each using the same techniques to form its results. The statistical equations and techniques that are put into the MOS products produce a wide range of weather information and forecasting predictors, ranging from temperature, dew point, cloud cover, wind speed and direction, precipitation types, amounts, and chances, visibility, and more. Although the data derived by the use of MOS is not foolproof, it is a great complement to other NWP models and is an important part of the forecaster's toolkit. A local forecaster can adjust the results for their specific location based on proximity to the MOS station, the localized terrain, as well as climatological history and other localized influences.

Two other major accomplishments of the 1970s can be wrapped into one, and that is the beginning of a more modern forecasting model. While the MOS was arguably one of the more important contributors to American meteorology at the time, the most important for European meteorology was arguably the creation of the European Centre for Medium-Range Weather Forecasts, or the ECMWF. This model first became operational in August of 1979, and has only grown in power and significance ever since (Lynch, 2007). The ECMWF is considered the first real-time medium-range forecast model, and produced forecasts five days a week, and eventually seven days a week by 1980 ("History," n.d.). The ECMWF model was the go-to computer model for Met Office weather forecasters during this time. It was created to limit the errors produced by the more simple models of the time, and tried to limit initialization errors as well as boundary errors that tended to present themselves the further out in a forecast time period. These boundary issues were due to weather features that were originally beyond the boundary of the grid, not being fully calculated and calibrated, creating an issue when they started to move into the grid. More detail is discussed in regards to these and other major computer models used today in Chapter 8.

Chapter Summary

Technology hit a major growth spurt starting around 1950 and kept growing through the 1970s, with television becoming a household item, and computers becoming a much-needed business tool. With the increase in television stations, news outlets, and the need to increase ratings, TV weathermen, and "weather girls," became staples of TV broadcasts, reporting first very primitively on black and white screens before evolving into professionalized forecasts in color and with more technical graphics. The forecasts produced on television would not have been possible without the boom in computer development and usage as well. While most televisions were in the United States and eventually Europe, computers were popping up in weather services across the world. With knowledge increasing about weather and the atmosphere along with growing mathematical and technological achievements, predicting the weather became more accurate, deeper into the future, and much more detailed, greatly surpassing anything that was seen before in the field of meteorology.

References

Book References

Edwards, P. (2010) *A Vast Machine: Computer Models, Climate Data, and the Politics of Global Warming*, Massachusetts Institute of Technology, Cambridge, MA.

Henson, R. (2010) *Weather on the Air: A History of Broadcast Meteorology*, American Meteorological Society, Boston.

Monmonier, M. (1999) *Air Apparent: How Meteorologists Learned to Map, Predict, and Dramatize Weather*, University of Chicago Press, Chicago.

Journal/Report References

Glahn, H.R. and Lowry, D.A. (1972) The use of Model Output Statistics (MOS) in objective weather forecasts. *Journal of Applied Meteorology*, 11.

Golding, B., Mylne, K., and Clark, P. (2004) The history and future of Numerical Weather Prediction in the Met Office. *Weather*, 59 (11), pp. 300.

Lynch, P. (2007) The origins of computer weather prediction and climate modeling. *Journal of Computational Physics*, 227 (7), pp. 3431–3444.

Persson, A. (2005) Early operational numerical weather prediction outside the USA: An historical introduction: Part II: Twenty countries around the world. *Meteorological Applications*, 12 (3), pp. 270–275.

UNESCO (1963) Statistics on radio and television 1950–1960. *Statistical Reports and Studies*.

Website References

"150 Years of Weather Forecasting for the Nation." (n.d.) *Met Office*, Met Office, http://www.metoffice.gov.uk/media/flash/s/s/MetOffice.swf, accessed 2 July 2015.

"History." (n.d.) *ECMWF*, European Centre for Medium-Range Weather Forecasts, http://www.ecmwf.int/en/about/who-we-are/history, accessed 2 July 2015.

"Leo Computers Society." (2015) *Leo Computers Society*, http://www.leo-computers.org.uk/index.html, accessed 2 July 2015.

"Moving Image Section: Motion Picture, Broadcasting and Recorded Sound Division." (n.d.) *American Women*, The Library of Congress, http://memory.loc.gov/ammem/

awhhtml/awmi10/television.html, accessed 2 July 2015.

Ravo, N. (1999) *Clint Youle, 83, Early Weatherman on TV*, The New York Times, http://www.nytimes.com/1999/07/31/us/clint-youle-83-early-weatherman-on-tv.html, accessed 2 July 2015.

"Supercomputers." (2014) *Met Office*, Met Office, http://www.metoffice.gov.uk/news/in-depth/supercomputers, accessed 2 July 2015.

Figure References – In Order of Appearance

"Historic Computer Images." (n.d.) US Army Photo, http://ftp.arl.army.mil/ftp/historic-computers/gif/eniac5.gif, accessed 18 July 2016.

"Robert FitzRoy and the Daily Weather Reports." (2016) Daily Weather reports. Met Office, http://www.metoffice.gov.uk/learning/library/archive-hidden-treasures/daily-weather-reports, accessed 10 July 2016.

5

Radar and Satellite History

Nicole Gallicchio

Technology began to revolutionize significantly after the 1950s. Computerized devices had emerged and uses for this system of technology rapidly grew. Meteorological advancements utilizing such technology began to take form, enhancing weather forecasting and comprehension of earth's atmosphere. These advancements derived from modifications of devices that were created for purposes outside meteorology. It was during World War II (1939–1945) and the Space Race (1955–1972) that this applied science developed, constructing the basis for modern technology.

5.1 Invention of the Radar

The invention of radar was not purposely designed for meteorological uses. It was for many years a carefully guarded US military secret, and many competitive interests became involved in its development and production before its information became public knowledge (Page Morris, 1962, p. 14). The term "radar" was coined as an acronym of radio detection and ranging, by two US naval officers. Radar is a technology that transmits and receives electromagnetic waves in order to perceive, trace, and gauge the speed of distant objects. Reflected waves are analyzed and processed by using mathematically derived equations,

specially designed and programed to overcome natural phenomena that affect the propagation of those waves.

The invention of radar was due to numerous previous scientific developments and findings, beginning with the development of radio waves. A German scientist, Heinrich Hertz (1857–1894), proved the existence of radio waves, by demonstrating the rapid variations of electric current. He experimented with these currents being projected into space, similar to those of heat and light, and it became known as a form of radio ("Heinrich Hertz," 2015). The concept of using radio waves was established by Nikola Tesla (1856–1943), an inventor, engineer, and physicist. Before World War II there were attempts to use CW (continuous wave) radio waves in order to detect the presence of ships; however, this system had flaws, enhancing the need for developing the technology. Not only was there a need for ship detection, but also there was a growing need to detect aircraft. Sir Robert Watson-Watt (1892–1973), a Scottish engineer, conducted experiments with radio waves. He was a meteorologist at the Royal Aircraft Factory and used his knowledge of radio to locate thunderstorms. In February 1935, Watson-Watt drafted a report titled *The Detection of Aircraft by Radio Methods*, which proved a success. These developments would see the

The Evolution of Meteorology: A Look into the Past, Present, and Future of Weather Forecasting, First Edition.
Kevin Anthony Teague and Nicole Gallicchio.

military application of radar during World War II ("Robert Watson-Watt," 2014).

Owing to the demand for radar use during the war, the advancements during the 1940s on radar uses outside the military were scarce and further development of radar was slow during this period. Meteorologists slowly began to pick up on and develop new knowledge and technologies regarding radio echoes. Radio echoes are electronic signals that are reflected back to the radar antenna. These signals proved valuable for scientists studying and identifying different atmospheric processes.

5.2 Development of Weather Radar

Weather radar may be considered the first of the remote sensing techniques used in meteorology (Figure 5.1). The use of radar for meteorological purposes developed as an outcome of the intense work being done during World War II on radar technology. Scientists began to notice that weather was causing echoes on their radar screens, and soon began experimenting with this phenomenon. Countries around the world at this time were creating their own experiments and projects, leading to extensive developments in meteorological radar. From all these further experiments, weather radar evolved and is now used as a major tool in weather forecasting.

There are several different distinctions of radar. Echo sounding is the most common arranged form used, in which radio waves are transmitted and received by a common antenna, and the radiation scattered back is in a form of an "echo." If the radar emits a continuous sine wave (CW radar) its range and frequency can only be used for certain tasks. Pulsed radar is the opposite to CW radar, and has proved the most effective way to use radar in meteorology. Pulsed radar works by the transmission of radiation in short-duration pulses with relatively long intervals of silence enabling the echo from one pulse to be received before the next is transmitted (Raghavan, 2003, p. 9). A Doppler shift is used when calculating the physics behind pulsed radar.

The early meteorological radars were used for two purposes. The first was to track balloons to determine upper-atmospheric winds and the second was the detection of precipitating cloud

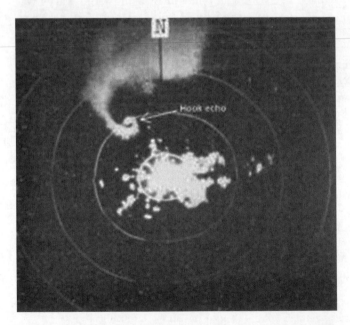

Figure 5.1 An early radar image of a thunderstorm and likely tornado in Meriden, Kansas, 19 May 1960. This image was taken from a WSR-3 radar, a lot different from the quality of imagery and information produced by today's technology. *Source*: Courtesy of NOAA and NWS.

systems (Raghavan, 2003, p. 3). Meteorological radar brought new uses in forecasting, allowing for a deeper understanding of atmospheric winds, clouds, and precipitation. Weather radars work by sending directional pulses of microwave radiation to sense objects and their motion. When determining precipitation, different colors will appear on the map depending on the intensity of the precipitation. Each color on the radar display will correspond to a different level of energy pulse reflected from precipitation ("How Radar Works," 2015). The availability to have visual analysis of live precipitation intensities provides meteorologists and scientists with current information regarding the atmosphere.

In 1949, radar was first used to obtain meteorological data as part of the US's Thunderstorm Project. This project was one of the first major field experiments in meteorology. During this experiment, thunderstorms were detected and monitored, using a large ground radar that was originally located near the operations area at the Clinton County Army Air Force Base until a longer-range, high-power radar was installed near Jamestown, Ohio. ("The Thunderstorm Project in Ohio," 2013). The density of observations used during the Thunderstorm Project had never been attempted before and set the standard for similar projects to come. The project demonstrated that radar could be used to detect the most dangerous parts of thunderstorms and guide airplanes around them. One of the most important findings of this project was the development of stages in the lifecycle of a thunderstorm ("The Thunderstorm Project in Ohio," 2013). The project became the foundation of future understandings and research in thunderstorms and weather phenomena.

In 1954, the Weather Bureau, Navy, Air Force, MIT's Institute for Advanced Study, and the University of Chicago formed a Joint Numerical Weather Prediction Unit at Suitland, Maryland. The first radar specifically designed for meteorological use was unveiled by the Air Weather Service, USAF. The radar was named AN/CPS-9

("Evolution of the National Weather Service," n.d.). The development of weather radar was a significant part of the advancements in meteorology and weather forecasting. Weather radar is a major part of today's technology in weather forecasting and storm warnings. Without the development of radar, mesoscale meteorology would not exist today. It was never feasible to connect precipitation patterns to synoptic-scale cyclones and fronts in a satisfactory way. Radar proved the ideal equipment for observing precipitation patterns (Raghavan, 2003, p. 107).

5.2.1 Doppler Radar

Doppler radar was invented by Christian Andreas Doppler, an Austrian physicist, who first observed that the frequency of light and sound waves can be affected by the relative motion of the source and the receiver. Doppler radar works by sending a beam of electromagnetic radiation waves at a precise frequency at a moving or stationary object. The electromagnetic waves are emitted, hitting the object and bouncing back toward the emitted receiver. Once the waves are transmitted back to the source, the results are interrupted using the relativistic Doppler effect (Zimmerman Jones, 2015). This works by calculating the change in frequency and wavelength of light that is caused by the relative motion of the source and the observer. When calculating this effect, one must take into consideration spatial relativity, which is the physical theory regarding the relationship between space and time. Doppler radar is a system that allows weather forecasters to have detailed analysis of weather patterns and their movements. Pulse Doppler radar is used to measure raindrop motion, precipitation, the structure of storms, and wind shears. Pulse Doppler works by transmitting pulses and receiving numerous samples ("What is Pulse Radar?" 2013). It determines the range to a target using pulse timing techniques and uses the Doppler shift of the returning signal to determine the intended object's velocity. Pulse radar was first used in the 1960s on fighter aircraft to allow them to detect other aircraft and objects. From a meteorological perspective, it enables wind speed to be determined, which is derived from the velocity of any precipitation in the atmosphere.

5.3 Invention of the Satellite

Before the Space Race, all instrumental readings and observations were acquired within earth's atmosphere. During World War II, some attempted aerial views above the atmosphere for use against the enemy. Experiments were done using cameras attached to rockets and retrieving the photographs after the rocket fell back down to earth. These methods proved impractical due to the expense and time restrictions inherent in using photographic data. As the need and curiosity grew to photograph earth from above, new technological devices were constructed. These devices became known as satellites, defined as objects that orbit planets that can be natural or artificial.

On 4 October 1957, the Soviet Union launched the first satellite into space, named Sputnik 1. Sputnik 1 was a sphere made of aluminum, measuring 58 cm (23 in) in diameter and weighing 83 kg (183 lb). It was powered by batteries and had four radio antennae that transmitted radio pulses into space (Owen, 2015, p. 8). It was placed into a low earth orbit, and its mission consisted of several goals. This was the first test, to see whether it was possible to launch a satellite into orbit around earth. The second was to be able to track its positioning. Being able to receive the data transmitted from the radio signals in space was an important goal of the overall mission. The mission was a success and the successful launch of Sputnik 1 sparked a new age of technology that introduced humans to life outside of earth's atmosphere. Although Sputnik 1 did not monitor weather on earth, the advancements and modifications made to the satellite were able to be directed toward meteorological purposes.

As technology became more advanced, the trajectory in which satellites orbit earth changed. A satellite stays in orbit because of the specific balance between its speed and the gravitational attraction of earth (Cobb, 2003, p. 17). They can inhabit one of three basic types of orbit: low, sun-synchronous, or geosynchronous orbit. Satellites also need to be stabilized once they are in place in order for them to stay in their orbit. They also require power to perform all the necessary functions. Early satellites were run by rocket engines and were battery powered. Today's modern technology has allowed them to be powered using solar power. Modifications to satellites were made progressively, enabling them to perform numerous tasks for multipurpose reasons.

5.4 Remote Sensing

The modern technology of remote sensing began with the invention of the camera in 1825. The ideas of photographing earth's surface from above emerged several years later, in the 1840s, when kites and balloons were being sent into the atmosphere. Still photographs were the only method of aerial views until the early 1960s. Remote sensing is the science of obtaining information about objects or areas from a distance, typically from aircraft or satellites ("What is Remote Sensing?" 2015). Remote sensing makes use of the interaction of energy waves to measure objects or materials from a distance, without physical contact. Remote-sensing weather equipment is used in the detection and measurement of weather phenomena. It can be used to predict weather patterns and track short-term and long-term weather trends, such as hurricanes and tropical storms. It is an essential technique for monitoring ozone depletion and global temperature fluctuations and trends (Blumenthal, 2015). The equipment used in the detection of meteorological elements is light energy through satellites, heat/infrared scanners on satellites, and radio waves through Doppler radar. This type of technology provides a way to observe large regions on earth ("Remote Sensing," 2014).

5.5 Development of Weather Satellites

The development of weather satellites derived from modifications to existing technologies. Once the first satellite was launched, meteorologists quickly realized this technology might be able to give scientists a better understanding of the atmosphere and its weather systems. "A weather satellite is a satellite that senses the state of the atmosphere by photographing cloud distribution, by using infrared photography to record cloud temperature as an indication of the height of the cloud tops, or by measuring microwave emissions from the atmospheric gases which can be translated into the atmospheric temperature at the height of the emissions" (Allaby, 2008, p. 630).

On 1 April 1960, the United States launched the world's first weather satellite into orbit, TIROS 1 (Thermal Infrared Observation Satellite). This satellite carried infrared cameras that could take still pictures and transmit them back to a receiver based on earth (Cobb, 2003, p. 27). TIROS 1 sent 4000 pictures a week back to earth for several weeks before its camera died. TIROS 1 was equipped with four transmitting antennae and one receiving antenna with a de-spin device. Two cameras were mounted in the baseplate parallel to the craft's spin axis. The cameras contained tiny vidicon tubes that were able to scan a 500-line picture in two seconds, which was then transmitted to the ground station (Hill, 1991, p. 10). These first images had numerous flaws, but it was a step toward new advancements and a better understanding of earth's atmosphere. Some of the problems meteorologists faced with these images consisted of the absence of discernible geographical features. Without the distinction between oceans and continents, meteorologists had no reference point as to what they were looking at. The only thing they could do was to compare the weather data they already had with the time stamp in which the satellite produced the image. Even without the reference points needed to be fully functional,

TIROS 1 began to give scientists an idea as to what weather systems and patterns actually looked like on a dynamic scale.

Weather satellites are usually launched into orbit on a booster rocket. Once the fuel is used up, the satellite will no longer function and it falls back to earth, where it is recovered. Once the weather satellite is launched above our atmosphere, like all other satellites, it stays in orbit thanks to the delicate balance between the speed and gravitational attraction to earth. Weather satellites collect a lot of data, which are sent back to the receiver on earth by radio, known as telemetry. Early weather satellites were spin-stabilized, the timing and control of the satellite cameras were not as precise as they are today, and pictures were also affected by the spinning of the camera. Today's weather satellites can function for several years. Once the satellite runs out of fuel or malfunctions, it remains in orbit, decreasing speed over time until the gravitational force pulls it down toward earth. The satellite decays over time and once it re-enters the atmosphere the friction begins to heat the satellite and eventually will cause it to burn up.

The development of weather satellites enabled the advancements of weather forecasting. They have become a crucial and necessary piece of the weather forecasting process. Ever since TIROS 1, numerous weather satellites have produced images of earth from above. For meteorological purposes this has allowed extremely detailed observations of cloud systems. Another major function of weather satellites has been the detection of tropical cyclones and hurricanes/typhoons. The creation of hurricane warning and watch systems has become a crucial part in making the public aware of threatening weather. With the images that satellites provide and the data that can be retrieved at the surface, meteorologists are able to piece together an understanding of weather elements and systems on earth from top to bottom, over land and water. By the mid-1960s, with the successful creation of a global picture of the surface and atmosphere of earth, the emphasis shifted toward measuring vertical distributions of temperature and moisture to better initialize global numerical weather prediction models ("History and Development of Satellite Meteorology/Climatology," 1998).

During the early 1970s, NASA developed two weather satellites that were placed into geosynchronous orbit, SMS-I and SMS-II. These weather satellites demonstrated the practicality of geosynchronous meteorological satellites, which led to the development of the Geostationary Operational Environmental Satellite (GOES) program. Since the emergence of this program, it has supported weather forecasting, severe storm tracking, and meteorology research in the United States. On 16 October 1975 the first satellite from the GOES program was launched, its name was GOES-1. In November 1979, during President Carter's administration, the National Oceanic and Atmospheric Administration (NOAA) was assigned management responsibility for all civil operational remote sensing from space. NOAA formed the National Environmental Satellite, Data, and Information Service (NESDIS) in 1982 with the merger of the National Earth Satellite Service (NESS) and the Environmental Data Service. Its purpose is to operate and manage the United States' environmental satellite programs, and manage the data and information gathered by the National Weather Service and other government agencies and departments. GOES satellite weather monitoring and forecasting operations give scientific researchers the use of data to better understand earth's land, atmosphere, ocean, and climate interactions. As of 2015, there are currently 15 GOES satellites that have been launched into space, with four more satellite launches anticipated between 2016 and 2025 ("NOAA's Geostationary and Polar-Orbiting Weather Satellites," 2014). (See Chapter 14, Section 14.4.2.)

Section Summary

The evolution of radar and satellites enhanced significantly as the decades went by. New modifications were made as new technology was developed. As more data were gathered and computerized devices were born, weather predictions became more accurate. Computer technology allowed for furthering operational models and systematic numerical weather processing. New mechanisms were added to satellites and updated existing devices to enhance images and data collected from the machines that circle earth's atmosphere from above. Remote sensing has brought a new dimension and understanding of the processes and elements that exist in our atmospheric system and also the impacts it has on humanity. All these technologies have become major ingredients in today's weather-forecasting processes. When combining satellite and radar data with computer-generated forecasting models, accuracy and dependability on weather forecasts began to soar. By 1980, a "weather man" or "weather girl" was not a comic relief on a news broadcast with drawn-on maps and black-and-white coloring but a daily necessity for the masses on how to prepare for the day, travel, and safety. TV stations dedicated solely to weather forecasting were to be developed as demand for media broadcast forecasting grew, and an explosion in computer technology led to much larger and more powerful computers to gather more and more information and perform more and more calculations, not only for organizations like the National Weather Service or the Met Office, but in the homes and the hands of the public as well.

References

Book References

Allaby, M. (2008) *Dictionary of Earth Sciences*, Oxford University Press, New York.

Cobb, A. (2003) *Weather Observation Satellites*, The Rosen Publishing Group, Inc. New York.

Hill, J. (1991) *Weather From Above*, Smithsonian Institute. Washington DC.

Owen, R. (2015). *Satellites*, The Rosen Publishing Group, Inc, New York.

Page Morris, R. (1962). *The Origin of Radar*, Anchor Books Doubleday & Company, Inc. Garden City, New York.

Raghavan, S. (2003) *Radar Meteorology*, Springer Science + Business Dordrecht, The Netherlands.

Journal/Report References

Blumenthal, R.L. (2015) Remote Sensing, *Salem Press Encyclopedia of Science*, January 2015.

Website References

"Evolution of the National Weather Service." (n.d.) *National Weather Service*. NWS and NOAA, http://www.erh.noaa.gov/gyx/timeline.html, accessed 25 October 2015.

"Heinrich Hertz." (2015) *Famous Scientists: The Art of Genius*, Famous Scientists, http://www.famousscientists.org/heinrich-hertz/, accessed 27 October 2015.

"History and Development of Satellite Meteorology/Climatology." (1998) *Satellite Climatology – GEOG 674*, University of Delaware, http://climate.geog.udel.edu/~tracyd/geog674/geog674_history.html, accessed 29 June 2015.

"How Radar Works." (2015) *Bureau of Meteorology*, Commonwealth of Australia, http://www.bom.gov.au/australia/radar/about/what_is_radar.shtml, accessed 18 June 2015.

"NOAA's Geostationary and Polar-Orbiting Weather Satellites." (2014) *NOAA SIS*, NOAA, http://noaasis.noaa.gov/NOAASIS/ml/genlsatl.html, accessed 24 October 2015.

"Remote Sensing." (2014) *JetStream: Online School for Weather*, NOAA & NWS, http://www.srh.noaa.gov/jetstream/remote/remote_intro.htm, accessed 15 June 2015.

"Robert Watson-Watt." (2014) *National Physical Laboratory*, NPL, http://www.npl.co.uk/people/robert-watson-watt, accessed 27 October 2015.

"The Thunderstorm Project in Ohio." (2013) *National Weather Service Forecast Office: Wilmington, OH*, NOAA & NWS, http://www.erh.noaa.gov/iln/research/ThunderstormProject/TSP.php, accessed 25 June 2015.

"What is Pulse Radar?" (2013) *InnovateUs: Innovation and Information for Sustainable Living*, InnovateUs, Inc.,www.innovateus.net/science/what-pulse-radar, accessed 1 June 2015.

"What is Remote Sensing?" (2015) *National Ocean Service*, NOAA, http://oceanservice.noaa.gov/facts/remotesensing.html, accessed 15 June 2015.

Zimmerman Jones, A. (2015) *How Does Doppler Radar Work?*, About Education, www.physics.about.com/od/physicsintherealworld/f/dopplerradar.htm, accessed 1 June 2015.

Figure References – In Order of Appearance

"1953 Beecher Tornado." (n.d.) *Beecher 50th Anniversary Commemoration*. NWS. NOAA, http://w2.weather.gov/dtx/beecherradar, accessed 20 July 2016.

Section III

Modern-Day Technologies, Advancements, and Social Media Impacts (1980–2013)

6

Personal Technology Boom

Nicole Gallicchio

As television and computerized devices became more advanced, weather outlets became vast. Personal technological devices were developed, creating new opportunities to further the broadcasting of weather forecasts. The creation of the Internet sparked one of the biggest booms of media weather coverage and personal forecasting abilities. Weather forecasting had become accessible to the public, and first-hand accounts were available to be viewed through advanced technological devices. Over the past 30 years accessibility to weather forecasting has become as simple as pressing a button.

6.1 Television Networks

The ability to broadcast weather forecasts and weather-related news to the public in a mass communication form proved itself a very effective method. Weather networks around the world began to take form, and weather-related news was regularly broadcast. The great success of weather-informing television broadcasts in countries such as the United States, the United Kingdom, Canada, and Japan influenced other countries to join the boom of weather broadcasting. International news broadcasts included weather segments in their routine schedules, often with weather-based or -themed lead stories and headlines concerning a weather phenomenon or an unfortunate disaster caused by extreme weather. This has become the same for local, regional, and all types of news agencies worldwide. Many news agencies often use the outlet of weather segments to increase ratings, especially since weather is a topic that everyone on earth can relate to, bringing uniformity to all cultures. Extreme weather has been a main topic for stories, debates, education, and political platforms from the 1980s to the present day. For popular news stations to stay relevant and retain viewers, they need their ratings to remain high, and weather news is a major contributor to this.

6.1.1 The Weather Channel

Launched on 2 May 1982, the Weather Channel became the first American weather-broadcasting network. The channel broadcasts weather forecasts, weather-related news, and documentaries. The Weather Channel originally gathered its national and regional forecasts from the National Oceanic and Atmospheric Administration (NOAA) and its local forecasts were sourced from the various National Weather Service Weather Forecast Offices around the country. Creating a national weather channel changed the way the American public viewed the weather as it was the first time weather-related information could be viewed around the clock.

The Evolution of Meteorology: A Look into the Past, Present, and Future of Weather Forecasting, First Edition.
Kevin Anthony Teague and Nicole Gallicchio.
© 2017 John Wiley & Sons Ltd. Published 2017 by John Wiley & Sons Ltd.

Over the years, the channel launched several different campaigns in order to keep the public interested and created new adaptations to the ever-changing technology. The Weather Channel knew that advertising would determine success or failure as they needed to appeal to their audience. According to Frank Batten, the channel's producers relied heavily on their gut instincts about consumers' weather information needs and about how much weather-based programming would appeal to them (Batten, 2002, pp. 49–50). There were major concerns that came along with sustaining a strictly weather forecast and news channel. Arguably, the biggest struggle the Weather Channel faced was whether viewers would sit through national weather forecasts to wait for their local forecasts. Local forecasts also created additional problems, with the technical side behind the comprehension of creating local forecasts across hundreds of local cable networks creating numerous challenges (Batten, 2002, p. 65). Another issue included the retrieval of all the data needed to have the local as well as national forecasts, with a common question being, "Where will all the data come from?" The main answer to that question was the National Weather Service (NWS) and its predecessor agencies. This did not come without its complications, however, because the channel still had to conform to specific rules that went with using the NWS data and warning systems. Originally, local data were retrieved this way, and national data were taken from the NOAA. WeatherStar (Weather Satellite Transponder Addressable Receiver) was created to generate local forecast segments on cable television systems (and eventually the Internet) nationwide. This program receives, generates, and inserts local weather forecasts and information, including advisories and warnings, into the Weather Channel's national programming system. WeatherStar enables forecasts to be more accurate and localized to a specific area.

Over the years, the Weather Channel has created several campaigns and slogans for promoting its network, modernizing its on-air presentations as technological advancements have presented themselves. In order to keep ratings up, adjustments were consistently made, along with adding programs that involved weather-related documentaries. Appealing to the public and providing useful weather information are ways the Weather Channel has been able to exist since its launch in 1982, beating all the odds and naysayers against a 24/7 all-weather television channel.

6.1.2 The BBC

The British Broadcasting Corporation (BBC) is the public-service broadcaster of the United Kingdom, headquartered at Broadcasting House in London. As stated in Chapter 3, Section 3.7, the first BBC weather forecast was a shipping forecast, broadcast on the radio on behalf of the Met Office on 14 November 1922. As of 23 March 1923, the corporation launched daily weather forecasts, with the first ever televised weather broadcasts in 1936, and since then has become the world's largest broadcasting center. The weather section of the BBC is so widespread that it can be viewed across the world.

Weather forecasts on BBC radio and television are recognized as the benchmark by which all other broadcasts in the United Kingdom and other countries around the world are judged. This traditional and renowned network has evolved rapidly with technology and has been at the forefront of weather distribution. The National BBC Weather Centre employs its meteorologists from the United Kingdom's Met Office. The purpose for this is the belief that the person telling the weather story should have a full understanding of the physical processes of the atmosphere (Giles, 2010, pp. 134–137).

6.1.3 World Networks

Global media and communication is an international effort to promote weather outreaches. Numerous countries air weather information and news on both a local and global scale. Outside of the Weather Channel and the BBC, there are numerous other networks around the world focusing on weather events, forecasts, and phenomena.

The Weather Network, a Canadian English weather news and information specialty channel, is partnered with the Weather Channel. Licensed by the Canadian Radio-Television and Telecommunications Commission on 1 December 1987, the Weather Network used computer-generated local forecasts that aired on the video feed of a live broadcaster. In 1994, the service began using the same system owned by the Weather Channel, WeatherStar. The Weather Network has progressed with technology in the twenty-first century, such as creating satellite services, Web and mobile services, HD broadcasts, and live programming.

Another world news network is Al Jazeera. Al Jazeera is a satellite television network, owned by the House of Thani, the ruling family of Qatar. The network is among the largest news organizations in the world, broadcasting throughout the Middle East and the rest of the world, providing weather news and information to a large part of the eastern world, and more recently the western world with what was called Al Jazeera America.

With television access becoming a worldwide normality, and with weather being one of the few things in life that affects everyone, albeit differently, broadcasting weather has become a priority and necessity. It is no longer just a simple, small, and forgotten segment on a localized one-hour news piece. Instead, weather broadcasting has gone straight to the forefront of all the major news outlets, local, regional, and internationally all across the world, further emphasizing the need to spread as much weather knowledge and information as possible to keep the world up to date and prepared.

6.2 Personal Computers

A personal computer (PC) is a digital computer designated for one person to use at a time. Computers that were small and inexpensive enough to be purchased and used in an individual's home became feasible during the late 1970s. The first personal computers manufactured on a grand scale were Apple Computer, Inc.'s (now Apple Inc.) Apple II, the Tandy Radio Shack TRS-80, and the Commodore Business Machines Personal Electronic Transactor (PET) (The Editors of Encyclopaedia Britannica, 2015). These computers were mainly used in secondary education schools and small businesses, having very little memory and were extremely slow. IBM Corporation came out with a rival computer in 1981 that was significantly faster and held more memory than any other PC on the market. During the mid-1980s, Apple introduced the Mac, which contained more graphically advanced features. As the 1990s approached, prices of PCs dropped and gained speed and functions. Laptops were also developed during this time and the development of portable computers was born. The end of the 1990s brought about the Internet and multiple Web browsers on personal computers in homes, schools, and businesses (see Section 6.3).

Computers replaced the telegraph for communicating information about weather, making it much faster and accurate. Personal computers in homes allowed the public to have access to information when they wanted it and how they wanted it. This also revolutionized the way meteorological forecasts were created, presented, and distributed. Not only did personal computers advance public knowledge but also the technological advancements aided meteorologists' and atmospheric scientists' accuracy and precision. The development of networks on personal computers allowed the sharing and transformation of weather information, observations, and data worldwide. With the cutting-edge technology in computers, processing meteorological data became the most efficient way to forecast the weather. Personal computers made this available to the public with new software and the development of the Internet.

6.3 The Internet and Meteorology

Arguably the most pronounced and biggest change for society – the invention of the Internet – opened up an infinite number of possibilities for the world, drastically changing society's day-to-day life. The invention of the telegraph, telephone, radio, and computer set the stage for this unprecedented integration of capabilities and the Internet revolutionized both computer technology and means of communication. The Internet is a worldwide broadcasting capability, a mechanism for information dissemination, and a medium for collaboration and interaction between individuals and their computers without regard for geographic location (Leiner *et al.*, 2015). The Internet dates back to the 1960s, when a computer scientist at MIT, J.C.R. Licklider (1915–1990), discussed his theories of a "galactic network." His theories led him to the head of computer research at the Defense Advanced Research Projects Agency (DARPA), where he convinced his successors of the importance of a networking concept. During the Cold War, technological advancements were necessary in case of nuclear attack. The United States needed a way to protect information should an attack arise. This eventually led to the formation of the Advanced Research Projects Agency Network (ARPANET), the network that ultimately evolved into what we now know as the Internet. ARPANET was a great success, but membership was limited to certain academic and research organizations who had contracts with the Defense Department. Research was continuously conducted from the late 1960s until 1983, when the Internet had become well established by a multitude of communities and daily computer communications. The first of January 1983 is considered the birth of the Internet. Prior to this, the various computer networks did not have a standard way to communicate with each other. A new communications protocol was established called Transfer Control Protocol/Internetwork Protocol (TCP/IP). This allowed different types of computers on different networks to "talk" to each other. ARPANET and the Defense Data Network officially changed to the TCP/IP standard on 1 January 1983, allowing all networks to be connected by a universal language ("A Brief History of the Internet," n.d.).

It wasn't until the end of 1990 that the Internet became accessible to the public via a subscription to a company such as AOL or CompuServe. Sir Tim Berners-Lee (1955–), a British computer scientist at CERN, a physics laboratory near Geneva, Switzerland, wrote the three fundamental technologies that remain the foundation of the Web; HTML, URL, and HTTP. Berners-Lee also wrote the first Web page editor/browser ("WorldWideWeb.app") and the first Web server ("httpd"). By 1991, people outside of CERN were invited to join this new Web community. The decision to make the underlying code available on a royalty-free basis was announced in April 1993, and sparked a global wave of creativity, collaboration, and innovation never seen before. In 2003, the companies developing new Web standards committed to a royalty-free policy for their work ("History of the Web," 2015). At this time the Web became open to all users with free access, with limitation being based on one's country.

At first, weather forecasts became readily available to subscribers with Internet access. They were able to see local and national forecasts provided by certain networks. Once the World Wide Web became free, anyone who had access to a computer had access to weather forecasts. This opened up a whole new realm of possibilities to the broadcasting and distribution of weather information and news.

6.4 Cellphones and Social Media

The first handheld cellphone was created by Martin Cooper and John F. Mitchell of Motorola in 1973. It weighed over 4 lb and was not commercially available. Its purpose was to make and receive phone calls via a radio connection while moving about a certain geographic area.

It wasn't until 1983 that Motorola came out with the first commercially available handheld mobile phone. Since that time mobile phones have grown into smartphones capable of Internet access, email, text messaging, and phone calls, and contain numerous applications that can be used for an infinite number of possibilities.

Mobile applications (apps) date back to the end of the twentieth century. They started out as games, calculators, and calendars. As time went on, consumers demanded more opportunities for their apps to aid in their daily lives. Meteorologists took advantage of this new method of providing information to consumers. Weather apps were developed and products were created in order to meet consumer demands. Modern technology allows for a variety of choices with regards to the various weather apps available on a smartphone. Most phones have built-in weather features, others need to be downloaded, some are free, and others need to be purchased. Weather-forecasting apps contain information such as temperature, wind speed, humidity, heat index, precipitation probability, and expected weather conditions for a period from 24 hours to a 5-day forecast, sometimes longer, depending on the application. Weather affects society's daily lifestyle and having the weather forecast at hand has become extremely valuable for consumers. For weather extremists and meteorologists, weather apps go further than just providing a computer-generated forecast. There are applications that contain live radar feeds, satellite data, and archive data. These different apps allow meteorologists to create their own forecasts and track the weather right from the palm of their hand. Weather applications have also enhanced warning systems, alerting people of impending or dangerous weather that is approaching their area. The Weather Channel has also taken part in the app business, creating its own application along with its sister company Weather Underground, which provides detailed weather-related news and forecasts. The simplicity that weather apps bring to daily life is unmatched by anything seen before in forecasting. People can plan their day or weeks by viewing their phone and have the weather forecast available to them in a matter of seconds.

Another aspect of cellphone technology is its connection with social media and networking. Social media can spark a chain link of news alerts that rapidly circulates information to the public. The number of connections that social media and networking allows is infinite. This creates an unlimited amount of access to information at the touch of a button on a smartphone or computer. "The first recognizable social media site, Six Degrees, was created in 1997. It enabled users to upload a profile and make friends with other users. In 1999, the first blogging websites became popular, creating a social media sensation that's still popular today" (Hendricks, 2013). After the invention of blogging, social media began to boom. In the early 2000s, sites such as Myspace and LinkedIn gained popularity and gave people a way to connect socially via the Internet. In 2005, YouTube was created, opening up a completely different genre of communication, allowing people to share and communicate via videos over great distances. In 2006, Facebook and Twitter were both launched worldwide, expanding the realm of social media and increasing the networking ability many times over.

A rapidly circulating social network allows meteorological news to be spread efficiently. There is no longer a wait for government announcements and local warning systems. The advantage of this method of awareness is the speed in which news and warnings are able to reach the public. A disadvantage is the authenticity of the information being distributed. Not all news being passed around the Internet and social media is factual or even credible. When dealing with weather-related information, corporations such as the NWS, the NOAA, the Met Office, the Weather Channel, or any private weather forecasting company, such as Forecasting Consultants LLC (there is more information on private forecasting companies in Chapter 10), would be the most effective in providing details and accuracy.

6.5 Movies

Advancements in the media allowed for society to become more aware of natural disasters and phenomena that occurred or were actively occurring in the world, and outlets such as television and the Internet enabled people to become more knowledgeable about earth's atmosphere and the way it reacts. Films were another way of disseminating information to viewers, although mostly as entertainment. The history of film began in the 1890s, with the invention of motion picture cameras and the establishment of film production companies. As new developments in motion picture cameras evolved the more realistic movies became. New technologies enabled filmmakers to take their subject matter to new levels. When overseeing movies as a whole, there has been a limited number of movies involving themes based upon meteorological elements and phenomena. Before the invention of computer-generated imagery (CGI), filming most meteorological happenings would have put film production workers in danger. Movies that contained events such as a tornado did so using special effects. For example, the tornado used in *The Wizard of Oz* (1939) was created from camera tricks and the use of a sock to replicate a tornado. With today's modern technology and an ever-growing interest in weather and its effects in the world, filming with segments including or fully about weather events have become much more common.

With the increase in filming technology, along with the increase in laptop computers and mobile phones, a new and dangerous recreation, research method, and occupation was created called "storm chasing" (discussed in detail in Chapter 7, Section 7.5). Storm chasing involves meteorologists, and weather enthusiasts, seeking to video any intense weather phenomenon possible. Movies can use this footage and insert it into their films, thus creating a very realistic effect. Documentary films are also very common and can be found on certain television networks. There have been numerous television shows based upon weather events as well, such as *Storm Chasers* and *Raging Planet*. These shows benefit the public because they bring a great awareness of earth's atmosphere and meteorological processes. They also bring a great deal of interest to viewers, initiating debates and curiosity for further learning about meteorology.

Arguably the most notorious weather film on record is *Twister* (1996). *Twister* is a fictional story that follows storm chasers trying to create an advanced warning system for tornadoes, risking their lives in the extremely dangerous path of tornadic thunderstorms. The film grossed worldwide nearly $500M ("Twister," n.d.). *Twister* was the first of its kind, sparking interest in storm chasing and advanced warning system technologies. The film also broadcast to the public images of tornadoes in ways many people were unaware of.

Another popular meteorology-themed film is the true story *The Perfect Storm* (2000), taken from the non-fiction book of the same title by Sebastian Junger. The story follows a fishing vessel that was lost at sea in the Perfect Storm of 1991. This storm was a nor'easter that absorbed nearby Hurricane Grace in the western Atlantic Ocean, moving up and off the coast of New England. In 2014, the film, *Into the Storm*, was released. This movie was fictional, telling the story of tornadic storms pushing through Oklahoma. The film received negative criticism, mainly regarding plot lines; however, the weather scenarios and visual effects received better reviews from the public. Whether the public enjoyed watching these films, they still provide semi-realistic views of natural weather disasters that can occur.

Climatological movies have been released that portray extreme meteorological phenomena. Such movies as *The Day After Tomorrow* (2004) and *2012* (2009) both have fictional storylines, portraying extreme climate change. These films grasp the attention of viewers because they show the extreme potentials of what earth's atmospheric processes could be capable of. Movies like these bring thrills and a subtle humbling of the strength of weather processes.

Chapter Summary

The importance of weather forecasting became evident as worldwide news agencies decided to make weather a top-selling news story on their networks. Weather also became the theme of its very own TV channel, the Weather Channel, further showing the value placed on weather, both locally and internationally. This coincided with the personal technology boom, which allowed for enhancements in the distribution and acquisition of weather forecasts and information. Personal computers, mobile phones, the Internet, and social media changed the face of communication, enabling availability to anyone with access to an infinite knowledge base. Meteorological data, news, and forecasts adapted accordingly to the technological progressions, improving communication between scientists and the public. Television, social media, and film also provided society with a more visual and interactive way to engage with and experience many different weather phenomena. With this added knowledge came a greater responsibility for meteorologists to more accurately forecast the weather and warn the public of impending weather situations.

References

Book References

Batten, F. (2002) *The Weather Channel*, Harvard Business School Press, Boston.

Journal/Report References

Giles, B. (2010) Weather broadcasting and training in the late twentieth century: The meteorologist's view. *Meteorological Applications: Special Issue: Communicating weather information and impacts*, 17 (2), pp. 134–137.

Website References

"A Brief History of the Internet." (n.d.) *Online Library Learning Center*, A project of the Board of Regents of the University System of Georgia, http://www.usg.edu/galileo/skills/unit07/internet07_02.phtml, accessed 24 July 2015.

Hendricks, D. (2013) *Complete History of Social Media: Then and Now*, Small Business Trends LLC, http://smallbiztrends.com/2013/05/the-complete-history-of-social-media-infographic.html, accessed 4 August 2015.

"History of the Web." (2015) *World Wide Web Foundation*, Creative Commons, http://webfoundation.org/about/vision/history-of-the-web/, accessed 4 August 2015.

Leiner, B.M., Cerf, V.G., Clark, D.C., *et al.* (2015) *Brief History of the Internet*, Internet Society, Creative Commons, http://www.internetsociety.org/internet/what-internet/history-internet/brief-history-internet#roberts, accessed 24 July 2015.

The Editors of Encyclopaedia Britannica. (2015) *Personal Computer (PC)*, Encyclopaedia Britannica, Encyclopaedia Britannica, Inc., http://www.britannica.com/technology/personal-computer, accessed 24 July 2015.

"Twister." (n.d.) *Box Office Mojo*. IMDb.com, Inc., http://www.boxofficemojo.com/movies/?id=twister.htm, accessed 4 August 2015.

7

Covering Major Storms

Nicole Gallicchio

Throughout the history of earth there have been storms that have been disastrous and life changing. News about these storms has always been passed along, whether it was the written or spoken word. As modern technology builds, coverage of these storms is continually evolving. Precautions and better predictions of weather are constantly evolving, allowing society to adapt appropriately to ever-changing weather conditions.

7.1   The Great Storm of 1987

The Great Storm of 1987 was a devastating weather event for Europe. Analysis of this storm suggests that its magnitude was unlike anything seen for centuries, and was a "once every few hundred years" type of storm. This storm began in the Bay of Biscay, south of the United Kingdom, on 15 October 1987, which is an unusual track for storms during the fall season. This storm may not have been the most powerful storm seen; however, it was so devastating because of the location it affected. The area's lack of preparation and experience with storms of this magnitude created a recipe for disaster.

The storm began due to warm tropical air and cold polar air colliding, ensuring the rise of warm air, creating a low-pressure system. The extreme differences in temperature produced

a very low pressure over the English Channel, reading 951 mbar at its lowest point (Figure 7.1). Due to atmospheric stability, air rapidly rose to the west of the low pressure, creating a significant difference in measurements. In order to balance the atmosphere, air flows from the high pressure to the low pressure, creating what we know as wind. The larger the pressure difference, the greater the wind speed. This storm along with cyclone force winds also had a sting jet. A sting jet is a surface wind maxima at the end of bent-back fronts. The sting jet is the culprit behind the maximum gust recorded during this event, clocking in at 164 mph (143 kts) ("How Did the 'Great Storm' of 1987 Develop?" 2012). The 1987 storm was also remarkable for the temperature changes that accompanied it. In a five-hour period, increases of more than 6 °C in an hour were recorded at many places south of a line from Dorset to Norfolk ("The Great Storm of 1987," 2015).

Weather forecasters knew that severe weather was on its way to the United Kingdom; however, computer weather models (see Chapter 8) indicated heavy rain would be the main component of this storm. Michael Fish, weather forecaster of the BBC, had reassured viewers that they would not be experiencing a hurricane, arguably leading viewers to take fewer precautions for the incoming storm. The areas affected by this storm were extremely ill prepared for the wind speeds and rainfall that ensued. In the early

The Evolution of Meteorology: A Look into the Past, Present, and Future of Weather Forecasting, First Edition.
Kevin Anthony Teague and Nicole Gallicchio.

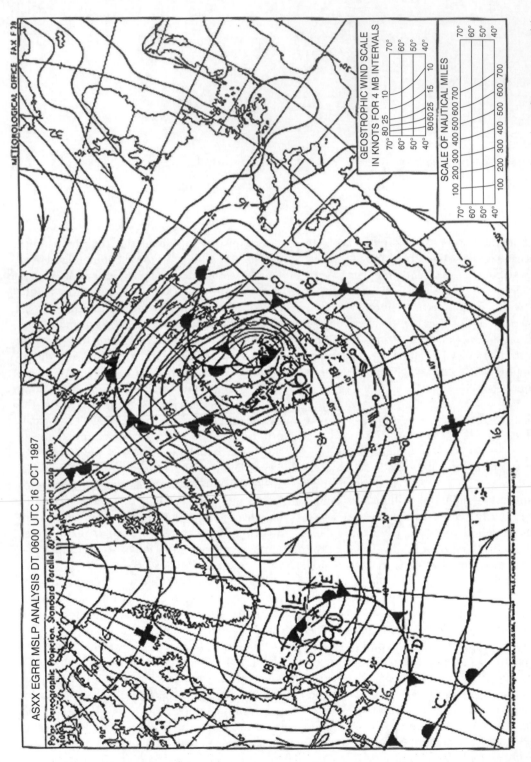

Figure 7.1 The Great Storm of 1987 is still remembered by those who rode it out in the United Kingdom. The weather map shows the early-morning model run on 16 October, showing a very strong low-pressure system on top of the United Kingdom. The tight isobars to the south and east of the low-pressure center show the locations of greatest pressure gradient and strongest wind speeds. *Source:* Image courtesy of © Crown Copyright 1987, under the terms of the Open Government License.

hours of 16 October 1987, the storm had progressed, and started to turn more north-east toward southeast England. The warnings of severe weather were issued to various agencies and emergency authorities, including the London Fire Brigade and Ministry of Defence ("Factfile: The 1987 Great Storm," 2015).

The storm made landfall in the early morning of 16 October 1987 in Cornwall, before tracking northeast. The strongest winds, however, were recorded along the southeastern edge of the storm, due to the sting jet, causing massive damage to areas in and around Sussex, Suffolk, and Kent. A total of 18 people lost their lives in Britain and four in France, totaling 22 people dead, with hundreds injured. Estimations of the cost of the damage were reported to be more than £1 billion. An estimated 15 million trees were destroyed, and roadways and railways were part of the wreckage. Thousands of homes were without power for several days ("Factfile: The 1987 Great Storm," 2015). This storm left a large path of destruction and has gone into the record books as The Great Storm of 1987, and as one of the largest forecasting mishaps in modern-day history.

7.2 Hurricanes and Tropical Cyclones

A "hurricane" is an intense storm of tropical origin, with sustained winds exceeding 64 kts (74 mph), which forms in the northern hemisphere over the warm Atlantic and eastern Pacific oceans (Ahrens, 2007, p. 406). This same type of storm is given different names in different parts of the world. In the northern hemisphere's western Pacific, the term "typhoon" is used. In and around India it is referred to as a "cyclone," and in and around Australia a "tropical cyclone." The international agreement is to refer to all hurricane-type storms that originate over tropical waters as "tropical cyclones" (Ahrens, 2007, p. 406). All mature tropical cyclones have common features, containing a clear calm center, known

as the "eye," that is bordered by a ring of extremely vigorous convection, known as the "eye wall." The extreme updrafts within the ring are fed by lines of converging winds that, in turn, form bands of thunderclouds and heavy rain (Fry *et al.*, 2010, p. 166). These spiraling storm systems can measure up to 800 km (500 miles) in diameter and can produce torrential rains, winds beyond 190 mph (165 kts), and extreme storm surges. When these storms make landfall they can be disastrous for anyone in their path. Some are notorious for the damage and causalities they cause; others, for their unusual behavior or meteorological extremes. As the population density continues to rise in coastal areas, many of which are vulnerable to hurricane tracks, the potential for large-scale devastation increases. Loss of life can be prevented if evacuations are instigated and accurate forecasting occurs; however, property damage is almost inevitable.

7.2.1 Hurricane Andrew

Hurricane Andrew was one of the most notable storms to ever make landfall in the United States, and was one of the most powerful storms to ever hit South Florida. Andrew hit the United States in August 1992, and was at that time the costliest hurricane in US history. The hurricane changed the landscape of South Florida completely and changed the way Floridians prepare for storms. Prior to Andrew, a very large hurricane in terms of size, Hurricane Betsy, struck Key Largo, Florida in August of 1965 and then entered the Gulf of Mexico. Betsy was a category 3 hurricane and was often used as an example for residents in South Florida when comparing storms. Andrew, on the other hand, was a very tightly wound hurricane, much smaller in size, but still greatly exceeded the damage done by Betsy.

Before Andrew made landfall, it had winds sustaining 165 mph (143 kts) in its eyewall, with wind gusts of approximately 175 mph (152 kts), which makes it a category 5 hurricane on the Saffir–Simpson hurricane wind scale, the strongest category given to hurricanes (Machos, n.d.). From 1900 until now, Hurricane Andrew is one of only three category 5 hurricanes to ever hit the United States. Its pressure in the eye of the storm measured 922 mbar, which at the time was the third lowest air pressure of any hurricane to make landfall in the United States (Landsea, 2002). Satellite pictures indicated that Hurricane Andrew formed from a tropical wave that crossed from the west coast of Africa to the tropical North Atlantic Ocean on 14 August 1992. "The wave moved westward at about 20 kts [23 mph], steered by a swift and deep easterly current on the south side of an area of high pressure. The wave passed to the south of the Cape Verde Islands on the following day. At that point, meteorologists at the National Hurricane Center, NHC, Tropical Satellite Analysis and Forecast, TSAF, unit and the Synoptic Analysis Branch, SAB, of the National Environmental Satellite Data and Information Service, NESDIS, found the wave sufficiently well-organized to begin classifying the intensity of the system using the Dvorak (1984) analysis technique" (Rappaport, 2005). For days after, convection subsequently occurred creating a cyclonic cloud rotation, becoming a tropical depression. The tropical depression began to grow, tracking into an environment in which strong southwesterly vertical wind shears and high surface pressures helped to create the evolution of this storm, reaching hurricane strength on 22 August 1992. By the time it reached Eleuthera in the Bahamas late on 23 August, it was a category 4 hurricane. After passing over the Bahamas, the hurricane moved westward toward South Florida. The strongest sustained wind occurred at the Fowey Rocks weather station with a recorded wind of 142 mph (123 kts), gusting to 169 mph (147 kts). The strongest wind gust during Andrew was reported at 211 mph (184 kts), in the northern eyewall. This measurement has been disputed and the wind speed is estimated to be closer to 178 mph (155 kts). After Hurricane Andrew crossed Florida, it entered the Gulf of Mexico, where it eventually made landfall in Louisiana

with wind speeds of up to 115 mph (100 kts) (Rappaport, 2005).

Hurricane warnings were issued in several states for this storm along with the Bahamas. Florida, Louisiana, and Texas were the major states affected by this storm. The damage done by Hurricane Andrew was due to extreme rain fall, very high and destructive wind speeds, and a large storm surge, and was estimated at around $25 billion. It is estimated that Hurricane Andrew directly cost 26 lives The additional indirect life lost brought the death toll to 65 people ("Introduction to Hurricanes: Science and Society," 2015). Many lives were saved by a combination of warnings, preparedness, and evacuation programs. However, many residents of South Florida at this time were inexperienced with hurricanes as several residents had just moved to the area right before the storm had struck. After the storm, many of these residents decided to leave and move to lower hurricane risk areas, while others decided to rebuild, a process that lasted for years after the storm.

Andrew raised a tremendous amount of awareness about hurricanes. It strengthened the precautions that Floridians take during storms. South Florida especially has continued to take its readiness to new levels, revising building codes and standards for safety during hurricanes.

7.2.2 Hurricanes Katrina and Sandy

Throughout history numerous hurricanes (tropical cyclones) have devastated regions. Some have become more notable than others for their meteorological properties or their destructive ways. Two other storms that are extremely notable in the United States are Hurricane Katrina and Sandy, both of which required severe relief efforts in the aftermath and caused record-breaking damage.

Hurricane Katrina is the costliest hurricane to date in the United States, with damage exceeding $100 billion. Katrina formed as a tropical depression on 23 August 2005. It made landfall in the Bahamas, Florida, and then eventually tracked into the Gulf of Mexico and slammed into Louisiana, affecting 145,000 km^2 (90,000 square miles) along the way. When Katrina hit the southern tip of Florida it was a category 1 hurricane. It then re-intensified and gained significant strength from the warm waters of the Gulf of Mexico. Katrina reached maximum wind speeds over the waters of the Gulf of Mexico on the morning of Sunday, 28 August of over 170 mph (150 kts, category 5), and its minimum central pressure dropped that afternoon to 902 mbar (Waple, 2005). Rainfalls exceeded well over 2.5 cm (1 in)/hour in some areas of the Gulf Coast. Excessive flooding was the product of low-lying land and broken levees in many parts of the coast, especially in New Orleans. Eighty percent of New Orleans was under water because of the broken levees and severe storm surge, leaving some parts under 6 m (20 ft) of water. The levees were designed to withstand a category 3 hurricane; however, Katrina was at category 5 strength for a period of time and only weakened to a high-end category 3 hurricane shortly before landfall. This led to a storm surge usually connected to a category 5 strength system to be pushed into the coast and into New Orleans, and the embankments could not hold the overflow of the water. Parts of Mississippi were also under 6–9 m (20–30 ft) of water due to the massive storm surges as well. It is estimated that 1836 lives were lost during the storm, with more than 1600 of these recorded in Louisiana, with hundreds of people unaccounted for and considered missing, creating a fluctuating death toll. The majority of deaths were caused by drowning, while other causes were injury, trauma, and heart conditions. Nearly half of the fatalities in Louisiana were senior citizens. There was also a lack of an effective evacuation plan in New Orleans for lower-income citizens who did not have personal transportation or the ability to use the limited public transportation within the city. Much debate and concern surrounded the mayor of Louisiana and the lack of accountability for the government money set aside to establish a workable evacuation plan. The response time and relief effort were also closely scrutinized worldwide. It took several

days for the Federal Emergency Management Agency (FEMA) to arrive with workers and proper paperwork and certificate requirements (Heldman, 2011). Tens of thousands of people who were evacuated ended up being left to fend for themselves in fairly poor conditions in hastily organized sites such as sports stadiums, without adequate food, water, or shelter. Recovery took several years, and cost billions of dollars, with some locations to this day still not having fully recovered financially or even structurally.

On 22–29 October 2012, Hurricane/Post-Tropical Cyclone Sandy moved from the Caribbean to the US eastern seaboard, ultimately making landfall near Brigantine, New Jersey, around 7.30 p.m. on 29 October. The storm had an enormous impact on life and property in both the Caribbean and the continental United States ("Service Assessment," 2013). This storm was unique in many ways. It was unprecedented for a storm track to approach coastal New Jersey, as Sandy did, from the east. Typically, tropical systems will track northward and often turning northeastward. The storm also merged with an intense low-pressure system that intensified the storm before it made landfall. The storm's formation, deformation, and transformation made it difficult to categorize and label. It began as a classic hurricane, gaining energy from warm waters of the Caribbean and moving toward the Gulf Stream. It then took a track toward the coast of New England, where it collided with a more classic fall/winter type storm system. The collision with a cold atmospheric air mass intensified the system and increased its size. This storm brought a storm surge that devastated much of the east coast, producing flooding that exceeded 2.5 m (8 ft) above sea level in some locations. The storm surge level measured at Battery Park, New York was 4.2 m (13.88 ft). New York Harbor's surf also recorded a record level wave of 9.9 m (32.5 ft) (Sharp, 2012). Coastal flooding occurred in Maryland, Delaware, New Jersey, and New York. Another unique factor is that Sandy produced blizzard conditions over the southern and central Appalachian mountain region, producing snow accumulations of over a foot in parts of West Virginia and North Carolina. The death toll from Sandy was estimated at 149. The estimated cost of property damage was $20 billion, and between $10–30 billion in lost business. Major differences from Hurricane Katrina were the preparations and the aftermath. Forecasts had predicted the massive storm, allowing proper precautions to be made and also the setting-up and planning of relief strategies. In August 2013, the Hurricane Sandy Rebuilding Strategy issued by the President's Sandy Recovery Task Force was implemented mandating that federal agencies work collaboratively across all levels of government and private sectors to approach suppleness ("Sandy Recovery Office," 2015).

7.2.3 The Great Bhola Cyclone of 1970

Sometimes tropical cyclones or other weather events play roles in history greater than just the destruction and devastation directly related to the storm itself. Oftentimes political issues result from natural disasters, similar to that of New Orleans from Hurricane Katrina, which have lasting effects. Even some storms that occurred prior to the modern-day weather era had catastrophic weather, political, and societal results. One example of an earlier storm that still has effects seen today occurred in 1970 in the densely populated country of Bangladesh, formerly East Pakistan. This area was struck by a tropical cyclone named the Great Bhola Cyclone, the deadliest tropical cyclone in recent history that reshaped national borders.

The cyclone formed over the central Bay of Bengal on 8 November 1970, traveling north and intensifying as it made its way toward East Pakistan (Bangladesh), reaching peak winds of 115 mph (100 kts) on 11 November 1970. The strength of the cyclone was equivalent to a category 3 hurricane. Meteorologists knew about the approaching storm; however, they had little to no way of warning those living on the coastal plain and on the islands that they were in the path of the storm. The Great Bhola Cyclone made landfall on the evening of 12 November 1970 in East Pakistan and eventually West Bengal (India).

On the night of landfall, there was an above-average lunar high tide. This high tide was accompanied by a 6 m (20 ft) storm surge and sustained winds of above 140 mph (122 kts), which together was beyond disastrous. Estimates of 300,000–500,000 lives were lost during the storm. The number of casualties was so high because of the time of day the storm stuck, and because of the lack of any warning. To date, the storm remains the deadliest tropical cyclone and one of the deadliest natural disasters in history. In the twentieth century, seven of the nine most deadly weather events in the world were tropical cyclones that struck the East Pakistan/Bangladesh region ("1970: The Great Bhola Cyclone," 2015).

There has been great criticism leveled at the Pakistani government for the relief operations following the disaster. The government acted extremely slowly with the relief efforts and supplies. The storm affected how the people viewed the government and contributed to the Indo-Pakistan War of 1971, splitting East Pakistan and creating the country of Bangladesh (Douglas, n.d.).

7.3 Tornadoes

A tornado is a rapidly rotating column of air that blows around a small area of intense low pressure with a circulation that reaches the ground (Ahrens, 2007, p. 389). Tornadoes, sometimes known as "twisters" or "cyclones," can assume a variety of shapes, sizes, and forms. The United States has the highest likelihood of tornadic activity; however, other parts of the world experience them too. Tornadoes occur where warm moist air can move at an angle to cool, dry air above. This typically will occur in the mid-latitude regions and on eastern sides of continents, often within locations where thunderstorms grow in the presence of wind shear (Fry *et al.*, 2010, p. 160). They can contain the strongest wind speeds on earth's surface, registering them as the most violent storms, and one of nature's most deadly natural phenomena. The development of predictability and warning systems can always be improved. This is mainly because tornadoes are the most difficult weather phenomenon to predict and monitor. Modern technology has had a difficult hold pinpointing tornadic recipes and patterns. In the 1960s, a man named Dr. T. Theodore Fujita (1920–1998), a Japanese American storm researcher at the University of Chicago, proposed a scale for classifying the rotating speed of tornadoes. It became known as the Fujita Scale. The Fujita Scale is still in place today, recently updated to the Enhanced Fujita Scale. A tornado's size is not necessarily directly correlated to its intensity and strength. There are also several categories and types of tornadoes, such as waterspouts, dust devils, and fire tornadoes.

7.3.1 Tornadic Outbreaks

When a large number of tornadoes form over a particular region in a small time span, it is referred to as a "tornado outbreak." The deadliest tornadoes are those that occur in families; different tornadoes that spawn from the same single, long-lived supercell thunderstorm (Ahrens, 2007, p. 390). These outbreaks are violent and cause massive amount of destruction. If a tornado outbreak is continuous or nearly continuous for several days, it is considered a tornado outbreak sequence. Tornado outbreaks can occur all over the world and at different times of the year; however, Tornado Alley in the United States sees the most tornadic activity than any other part of the world. Tornado Alley mainly refers to the southern central portion of the United States. It is in this region that the proper atmospheric setting for the development of severe thunderstorms is ideal. Here warm, humid air at the surface is blanketed by cool, dry air aloft, which creates a conditionally unstable atmosphere. When a strong vertical wind shear is in place, which is usually provided in this area by the polar jet stream, the surface air is forced upward, causing large thunderstorms to develop and tornadoes to be produced.

From 3 to 4 April 1974, the second largest tornado outbreak to date occurred. This outbreak is referred to as the "Super Outbreak" (Figure 7.2). Within a 24-hour period, 147 tornadoes were confirmed throughout the United States, from Alabama northward into Michigan, and eastward into Virginia. Thirteen states were struck by tornadoes in the United States. The entire outbreak caused over $600 million in damage in the United States and caused 330 casualties ("1974 Tornado Outbreak: The Worst in US History," n.d.). The largest outbreak to date occurred on 25–28 April 2011. The current count of strong tornadoes for the April 2011 outbreak is three EF-5, 12 EF-4, and 21 EF-3 ("EF" standing for the tornado wind speed and strength scale called the Enhanced Fujita Scale – EF-0 the weakest through EF-5 the strongest). In comparison, the April 1974 Super Outbreak had six F-5s, 23 F-4s, and 35 F-3s ("F" standing for the Fujita Scale – F-0 through F-5 – prior to the updated and enhanced distinguishing of categories) ("Tornadoes: April 2011," 2011).

As previously mentioned, the country of Bangladesh is notable for massively devastating results after storms and tornadoes, due to the lack of safe shelter and an extremely dense population. On 26 April 1989, a tornado ripped across Manikganj District, Bangladesh, and killed approximately 1300 people and left over 80,000 homeless. The tornado left an eight-mile path across the country, devastating the lives of tens of thousands of people (Fry *et al.*, 2010, p. 162).

7.3.2 Advanced Warnings for Tornado Activity

Atmospheric scientists and meteorologists monitor thunderstorms and the stability of the atmosphere in order to better predict the occurrence of tornadoes. If tornadoes are likely to form, a tornado watch is issued hours beforehand. In the United States there is the Storms Prediction Center, which is able to alert the public about potential tornadic activity within a certain region and timespan. Once a tornado is spotted by humans or radar in the United States,

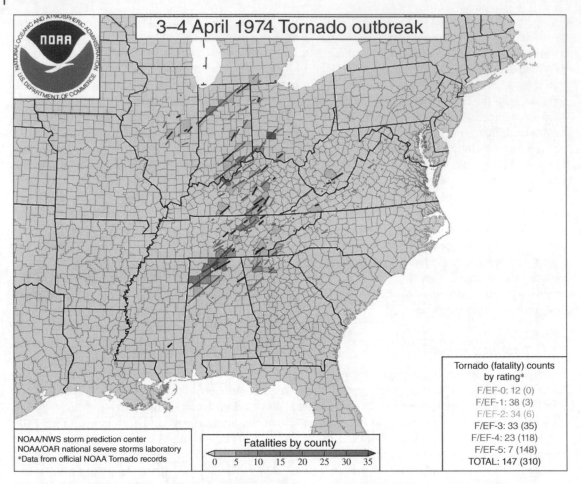

3–4 April 1974 Tornado outbreak

NOAA/NWS storm prediction center
NOAA/OAR national severe storms laboratory
*Data from official NOAA Tornado records

Fatalities by county
0   5   10   15   20   25   30   35

Tornado (fatality) counts
by rating*
F/EF-0: 12 (0)
F/EF-1: 38 (3)
F/EF-2: 34 (6)
F/EF-3: 33 (35)
F/EF-4: 23 (118)
F/EF-5: 7 (148)
TOTAL: 147 (310)

Figure 7.2 The map shows the locations and distances of the tornado outbreak of 3–4 April 1974 in the United States. The table in the bottom right corner shows the total number of each category of tornado, along with the fatalities that occurred in each category tornado. In total, 147 tornadoes occurred, resulting in 310 deaths. Tornadoes spanned from Alabama northward into Michigan, and as far east as central Virginia. *Source*: Courtesy of NOAA. (*See color plate section for the color representation of this figure.*)

a tornado warning is issued by that area's local National Weather Service office. Across the globe, different countries have different methods of alerting the public of severe weather. Many countries have specific meteorological societies that issue warnings regarding thunderstorms and tornadoes. Other forms of communication regarding weather warnings, besides television, include radio, weather radio, siren, and more recently cellphone alerts. Seeking proper shelter during a tornado is crucial to survival and these warnings are the only way to stay safe.

7.4 Floods, Droughts, and Wildfires

Many natural disasters are not due to weather elements in the atmosphere but are caused and directly influenced by earth's weather processes. Floods, droughts, and wildfires are all ignited and fueled by weather processes such as wind, rain, and temperature. These natural phenomena can cause damage to earth's atmosphere and climate, along with loss of life and devastation to communities.

On earth, something is always burning. All over the globe wildfires are common phenomena in arid regions, often caused by lightning strikes or natural influences, and sometimes by human negligence or arson. Fires can generate significant amounts of smoke pollution, release greenhouse gases, and destroy ecosystems. They emit large quantities of carbon monoxide and dioxide, soot, and other aerosols that affect earth's atmospheric composition. Global patterns of wildfires have developed over time, due to the natural cycles of rainfall, dryness, and lightning ("Fire," 2000). According to NASA's (National Aeronautics and Space Administration) global maps, cyclical fires predominately occur in South America and Africa seasonally. Some wildfires are even caused purposely by humans to restore and manage farmlands and clearings for natural vegetation restoration, which can be good for an ecosystem. The weather's effect on wildfires can directly influence the ignition, expansion, and extinguishing of the blaze. Heatwaves, droughts, winds, and cyclical weather and climate patterns can directly impact the behavior of wildfires. Fires can devastate populated areas, causing loss of life and infrastructure.

Drought is an insidious natural hazard with far-reaching impacts. It is the oddest form of extreme weather because it does not come with a strong known presence. Drought develops slowly over time and occurs all over the globe. It cannot merely be defined as low rainfall. A meteorological drought is in relation to the amount of precipitation over an area. However, there are also hydrological droughts and agricultural droughts, referring to the effects of precipitation on water levels, soil moisture, groundwater, and crop water demands. Both weather and climate are directly connected to drought. Rain is unevenly distributed through out the globe and some areas will always naturally receive less rainfall than others (Fry *et al.*, 2010, p. 186). Therefore, the term "drought" remains a relative term and depends on the climate for a specific area. For example, in the

United Kingdom a drought is considered after a period of 15 days without precipitation. However, in Libya a drought is classified if no precipitation has fallen within two years (Fry *et al.*, 2010, p. 186). Drought can threaten drinking water, food, and ecosystems. Water conservation around the globe is an ever-present and ever-growing issue, and efforts to limit the effects that drought has on society and ecosystems are ongoing. Technological advancements in the twenty-first century have reduced water use through the employment of more-efficient appliances and irrigation technologies. Awareness of recycling and reuse of water is expressed all over the globe, with agencies such as the Red Cross, Global Waters, and the United Nations addressing and educating about better water management.

Flooding is the exact opposite of drought. It is an excess of water that submerges land that is typically dry. Flooding can develop in many different ways, through meteorological processes and atmospheric phenomena. Flash floods can occur within a few minutes or hours of excessive rainfall, a dam or levee failure, storm surge, or a sudden release of water held by an ice jam. Some floods can develop very slowly without any visible signs or rain. Overland flooding, the most common type of flooding event, typically occurs when waterways such as rivers or streams overflow their banks as a result of rainwater, snowmelt, or a possible levee breach causing flooding in surrounding areas. It can also occur when rainfall or snowmelt exceeds the capacity of underground pipes, or the capacity of streets and drains designed to carry floodwater away from urban areas ("Floods," 2015). Coastline flooding is a quickly developing area of interest as discussions of climate change and its effects arise. Approximately 40% of the world's population lives within 100 km (60 miles) of a coastline (Fry *et al.*, 2010, p. 424). Most of the coastlines around the world are either at or below sea level, or in low-lying lands. These communities are at a greater risk for flooding, increasing the possibility of disastrous outcomes. Coastline protection against flooding has been introduced and implemented already by several nations; however, without the proper calculations and adaptations, these boundaries may not hold.

7.5 Storm Chasing

Even dating back to the era when humans solely believed in folklore, they have always been fascinated with weather phenomena. Tracking, recording, understanding, and monitoring the weather have been going on since ancient times. The pursuit of severe weather conditions has become known as "storm chasing." There are many different motives behind storm chasing. Some seek the adventure and thrill, while others seek scientific research and media coverage. Many chasers are using technology such as photography and videoing for a multitude of reasons. There are websites formed by storm chasers on which you can view live storm chasing videos and have access to radar to aid in tracking storms. Forecasting and intercepting a storm brings challenges and thrills. This goal of capturing and learning up close and personal with severe weather poses extreme risks. The predictability of storms on a momentary basis is unprecedented, leaving unknown factors and dangerous scenarios threatening the lives of even the most experienced storm chasers. Recreational storm chasing is advertised and can be part of a vacation tour with many companies offering the thrill of storm chasing for a price, primarily in the United States in Tornado Alley, due to the high volume of storms creating a high probability of severe weather intercepts.

This dangerous pursuit of severe weather conditions can also provide potential scientific research. For the scientists that storm chase, the ultimate goal is a better understanding of how storms operate in order to increase their predictability. Other than science, storm chasing brings adventure, and videos can be used in the media and film industry for a variety of reasons. Many times storm chasers end up becoming first responders to devastation. They typically are so closely tracking a storm that they are the first to enter an area that has been hit. This can save lives, due to their fast response time. Knowledge of the way storms track can also save lives through prevention and seeking proper shelter when alerted by a passing storm chaser. Sometimes a storm can develop so quickly that it is difficult to have proper warning in such a small amount of time. These storm chasers are usually close to the scene of a storm, able to provide information and warning to the public as well as the local National Weather Service offices.

Storm chasing continues to intrigue many people, mainly because of the excitement and sense of thrill-seeking it involves. The crucial scientific research and data that have been and continue to be gathered and calculated by the research-based storm chaser have proved vital to our understanding of the inner workings and development of tornadoes. Chasing severe weather, and enhancing our knowledge thereby, will continue, just as it has since ancient times.

7.6 Media Coverage

The most massive form of weather communication is via media coverage. The news provides imperative information to the public on all topics. Weather coverage is of great importance due to the threat it poses to life and property damage. There are many different controversies, debating the pros and cons of media coverage, especially when presenting the weather. Many meteorologists feel the media has produced a negative image of weather forecasting, creating doubt in the accuracy and seriousness that should be taken by the public regarding severe weather. The thoughts, influences, and activity that prevail over the public via mass media create an alternate viewpoint.

Media coverage of the weather allows preparations to be made involving weather protocols. Emergency announcements are made via the media along with updated forecasts. The public relies heavily on the announcements and information provided by the media as the main sources of reliable scientific information that many people have access to. Not everyone has received a thorough education in meteorological forecasting. The media providing and explaining the weather forecast, by showing satellite and radar images, helps to make the public more knowledgeable on the topic. Seeing weather data and their interpretation increases the public's understanding and so improves its decision-making regarding the weather.

Many representatives of the news media report on the aftermaths of weather phenomena, advertently becoming directly involved with the unfolding events. Media coverage of the products of severe weather brings knowledge and awareness to those both being affected and those who are outsiders. Media coverage can help spread the word on disasters and can allow people to see the damage and destruction done by different types of weather phenomena. With knowledge of the potential damage comes the likelihood of more preparations being made. When people see the potential effects of severe weather, they can typically visualize what would happen if severe weather were to strike their area. Footage of disaster zones can aid in funding what is needed in order to rebuild, restore, and provide sufficient supplies to the affected communities. Typically there is a lack of manpower when rebuilding, and volunteers are always needed in restoration projects after destruction occurs. Media coverage allows large numbers of people to receive word quickly that help is needed in devastated areas.

While there are seemingly many pros to media coverage on weather systems, such as education, warnings, and aftermath relief advertising, there can also be several cons. The question always arises of when does media coverage cease to be news and start to become hype? This line can be very thin and difficult to determine. Lethal storm coverage is essential in saving lives; however, the market for hype can mean doubt ensues and lead to a lack of preparation for these storms. Due to inaccuracies in weather forecasts, many tend to believe that there is a level of exaggeration that frequently takes place in the news reporting the forecasts, especially in a severe weather forecast. Accuracy and awareness of storms in the past hundred years have been lacking. This evidently left reasonable doubt and skepticism about forecasts, which has lingered with the public. However, as technology advances and forecasting becomes more accurate, the emphasis on the severity of storms portrayed in the media may result in a better interpretation of warning systems. There has been a rise in viewers interested in severe weather and its aftermath within the last 20 to 30 years as people are more interested in seeing the media coverage of satellite and radar images, along with explanations of what is transpiring.

Chapter Summary

Throughout history, weather has shaped society, geography, politics, and livelihoods. Certain extreme weather events, such as Hurricane Katrina or The Great Bhola Cyclone, are prime examples of the effects storms can have through the weather itself and through its aftermath. Tornado outbreaks, although on a smaller scale, bring with them earth's most violent weather conditions. The chasing of storms has become a way to protect, learn, and entertain the masses. There are numerous ways that storms and severe weather have been broadcast and portrayed through the media. Media coverage enables a massive amount of communication that is necessary in order to warn the public of imminent weather, creating better awareness and aid for disaster zones, as well as educating society about meteorology.

References

Book References

Ahrens, C.D. (2007) *Meteorology Today: An Introduction to Weather, Climate, and the Environment.* Eighth edition, Thomson Brooks/ Cole, Thomson Higher Education, Belmont, CA.

Fry, J.L., Graf, H.F., Grotjah, R., *et al.* (2010) *The Encyclopedia of Weather and Climate Change*, University of California Press, Los Angeles.

Journal/Report References

"Service Assessment," (2013) Hurricane/Post-Tropical Cyclone Sandy, October 22–29, 2012. United States Department of Commerce, NOAA, Silver Spring, MD.

Waple, A. (2005) Hurricane Katrina, NOAA's National Climatic Data Center, Asheville, NC.

Website References

"1970: The Great Bhola Cyclone." (2015) *Hurricanes: Science and Society*, The University of Rhode Island, http://www.hurricanescience.org/history/storms/1970s/greatbhola/, accessed 11 August 2015.

"1974 Tornado Outbreak: The Worst in US History." (n.d.) *Public Affairs*, NOAA, http://www.publicaffairs.noaa.gov/storms/description.html, accessed 12 August 2015.

Douglas, R. (n.d.) *Bhola at a Glance*. DisasterProject10, http://disasterproject10.wikispaces.com/Bhola+Cyclone+Facts, accessed 11 August 2015.

"Factfile: The 1987 Great Storm." (2015) *Met Office*, Met Office, http://www.metoffice.gov.uk/news/in-depth/1987-great-storm/fact-file, accessed 10 August 2015.

"Fire." (2000) *Global Maps*, NASA Earth Observatory, http://earthobservatory.nasa.gov/GlobalMaps/view.php?d1=MOD14A1_M_FIRE, accessed 17 August 2015.

"Floods." (2015) *Ready: Prepare, Plan, Stay Informed*, Ready, http://www.ready.gov/floods, accessed 17 August 2015.

Heldman, C. (2011) *Hurricane Katrina and the Demographics of Death*, The Society Pages: Sociological Images, http://thesocietypages.org/socimages/2011/08/29/hurricane-katrina-and-the-demographics-of-death/, accessed 16 August 2015.

"How Did the 'Great Storm' of 1987 Develop?" (2012) *Met Office News Blog*, Met Office, http://blog.metoffice.gov.uk/2012/10/12/how-did-the-great-storm-of-1987-develop/, accessed 6 August 2015.

"Introduction to Hurricanes: Science and Society." (2015) *Hurricanes: Science and Society*, University of Rhode Island, www.hurricanescience.org, accessed 27 October 2015.

Landsea, C. (2002) *Re-Analysis Project: Hurricane Andrew's Upgrade*, National Hurricane Research Division, NHC, NOAA, http://www.aoml.noaa.gov/hrd/hurdat/andrew.html, accessed 27 October 2015.

Machos, G. (n.d.) *The Story of Hurricane Andrew*, Hurricaneville.com, http://www.hurricaneville.com/andrew.html, accessed 10 August 2015.

Rappaport, E. (2005) *Preliminary Report: Hurricane Andrew 16–28 August, 1992*, Addendum: Category 5 upgrade, NCEP, NHC, and NOAA, http://www.nhc.noaa.gov/1992andrew.html, accessed 10 August 2015.

"Sandy Recovery Office." (2015) *Federal Emergency Management Agency (FEMA)*, Department of Homeland Security, https://www.fema.gov/sandy-recovery-office, accessed 16 August 2015.

Sharp, T. (2012) *Superstorm Sandy: Facts about the Frankenstorm*, LiveScience, http://www.livescience.com/24380-hurricane-sandy-status-data.html, accessed 13 August 2015.

"The Great Storm of 1987." (2015) *Met Office*, Met Office, http://www.metoffice.gov.uk/learning/learn-about-the-weather/weather-phenomena/case-studies/great- storm, accessed 10 August 2015.

"Tornadoes: April 2011." (2011) *National Centers for Environmental Information, NOAA*, Department of Commerce, https://www.ncdc.noaa.gov/sotc/tornadoes/201104#0426, accessed 12 August 2015.

Figure References – In Order of Appearance

Met Office Press Office. (2012) How did the "Great Storm" of 1987 develop? Met Office, https://blog.metoffice.gov.uk/2012/10/12/how-did-the-great-storm-of-1987-develop, accessed 25 June 2016.

Prociv, K. (2013) Looking back at the April 3–4, 1974 Super Outbreak. US Tornadoes. Tornado History, http://www.ustornadoes.com/2013/04/03/looking-back-at-the-april-3-4-1974-super-outbreak, accessed 12 July 2016.

8

The Rise of Modern Computer Models

Kevin Anthony Teague

As stated in Chapter 4, computer models began to revolutionize the way meteorologists were able to develop their forecasts and increase their skill score. Through the 1950s, 1960s, and 1970s, computer models were very basic, not allowing for too much information nor calculations and limiting the accuracy and duration of forecasts. As the sophistication of computer technology grew, computer power and capacity increased as well, allowing for larger and more powerful computer models to develop and take over the forecasting world.

8.1 European Center for Medium-Range Weather Forecasting (ECMWF)

The ECMWF has arguably been the pioneer and leader of all modern-day forecasting models, dating back to 1979 (Figure 8.1). Since then, at least one 10-day forecast has been generated each day. This model is not used solely for the purposes of media-style forecasts, or for public use. Instead, the ECMWF has been used as a vital resource for many of the world's top meteorological and space agencies and departments. As stated in the ECMWF cooperate brochure in 2012, "In 3 decades ECMWF has improved accuracy and reliability of weather forecasting, working in collaboration with member and cooperating states, the European Union, and partners such as the World Meteorological Organization (WMO), the European Organization for the Exploitation of Meteorological Satellites (EUMETSAT), and the European Space Agency (ESA)" (ECMWF, 2012, p. 2).

Many other beneficiaries of this model (as well as other models to be discussed) include national hydro-meteorological services, scientists monitoring environmental and climate changes, military, local authorities, emergency departments, energy providers, shipping agencies, commercial fishing, weather-sensitive manufacturing companies, transportation systems, policymakers, and general meteorological forecasting providers. The data gathered, analyzed, and dispersed from this and most other models come from a wide range of technologies, totaling over 300 million observational data instruments per day. Some of these involve earth observing systems, satellites, radar, automatic and manned weather stations, aircraft, ships, weather balloons, and buoys (ECMWF, 2012, p. 4). Taken together, this information is calculated and generated into five components: (1) general circulation model, (2) ocean wave model, (3) data assimilation system, (4) ensemble prediction system, and (5) monthly forecasting system (Persson and Grazzini, 2007, pp. 3–6).

The Evolution of Meteorology: A Look into the Past, Present, and Future of Weather Forecasting, First Edition.
Kevin Anthony Teague and Nicole Gallicchio.

Member states of the ECMWF

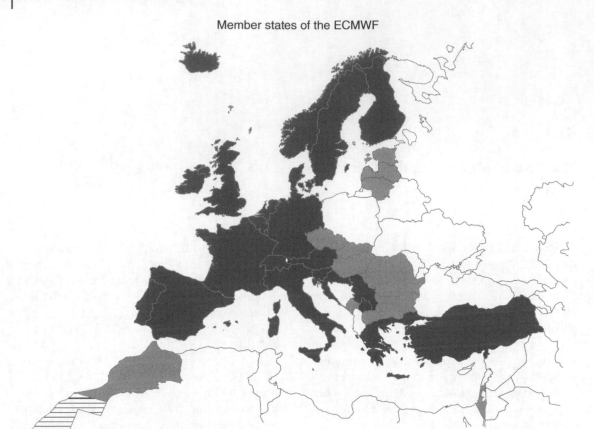

Figure 8.1 The European Centre for Medium-Range Weather Forecasts (ECMWF) is a research institute as well as an operational weather service. The Member States which provide to and receive from the ECMWF consist of most of Europe, Israel, and parts of northern Africa. In total, there are 22 Member States (dark shading), and 12 Cooperating States (lighter shading and dashed shading). In addition, the ECMWF also has formal cooperation agreements with many non-Member States and international organizations. *Source*: Reproduced with permission of ECMWF.

8.1.1   General Circulation Model

The original ECMWF was developed as a grid-point model with 15 levels going as high up as 10 hPa, with a horizontal grid length of 200 km (124 miles). Vertical levels in computer models represents a scale system measured in higher air pressure toward the surface of earth, the bottom level, and the lower pressures further up in the atmosphere, the top level. The more vertical levels a model has, the better the all-round picture of the atmosphere, from top to bottom, the model will have, therefore theoretically creating a better forecast and prediction of atmospheric conditions. By 1983, the grid-point style changed to a model with spectral representation, and consisted of a triangular truncation at wave number 63. The higher the wave number, the stronger and more uniform the resolution. In 1985, the spectral resolution increased to 106 with 19 vertical levels. The next advancements occurred in 1991, when the spectral resolution increased to 213 along with an increase in vertical levels to 31. By 1995, a cloud scheme was introduced. This improved forecasts involving cloud development and prediction. In return, this also greatly increased the skill of forecasting precipitation amounts and locations. All together, this advancement greatly impacted the model dynamics as a whole. By 1998, horizontal spectral resolution was at 319, and the vertical levels were up to 50. Another advancement occurred in 1999, and was to include ozone as a predicted variable. By 2000, spectral resolution increased to 511, and vertical levels to 60. The constant upgrading continued into 2006, when spectral resolution and vertical level resolution increased to 799 and 91 respectively. These increases continue today (with vertical levels now sitting at 137) and will do so into the future as more and more advancements are discovered and implemented into the computer models (Persson and Grazzini, 2007, pp. 3–4).

8.1.2 Ocean Wave Model

By 1992, an operation model consisting of a portion of the North Atlantic and European waters was implemented. This was followed by a wave model that was introduced in 1998 and was integrated into the ECMWF, allowing for the analysis of wind and waves over the open waters. This advancement was also used in the analysis of ensemble, seasonal, and monthly systems (Persson and Grazzini, 2007, p. 4).

8.1.3 Data Assimilation System

In 1991, a one-dimensional variational scheme was introduced to include satellite radiance assimilation into the model. A three-dimensional scheme was introduced in 1996 to include all types of observations. This was quickly upgraded to a four-dimensional scheme in 1997. These developments finally started to catch up to the development of satellite power. With these new four-dimensional schemes, more and more of the extremely complex information stemming from data collected by satellite was assimilated into the model. These data were always available, because of satellites, but were never able to be fully taken advantage of inside the models until the upgrade to the four-dimensional schemes. Meteorological data were most heavily based on radiosonde data, but today the data involved in these schemes are at the forefront of forecasters' and researchers' work.

Now, infrared and microwave radiances can be included into the ECMWF, increasing skill in forecasting temperatures and humidity analysis (Persson and Grazzini, 2007, pp. 4–5).

8.1.4 Ensemble Prediction System (EPS)

An ensemble model consists of a variety of model runs, each with a different initialization of data and each with different small perturbations within each model run. These differences account for many of the possibilities that may occur in the atmosphere through the forecast cycle, showing the tops and bottoms of the limits of uncertainty in a forecast. The more a model runs with different variables, the better the representation of possible outcomes. The ensemble then takes the mean of all the runs and displays an average of all the variations. The Ensemble Prediction System (EPS) of the ECMWF first consisted of 32 members with a spectral resolution of 63 and with 31 vertical levels in 1992. Advancements in the amount of vertical levels occurred in 1996, increasing to 50 levels. A wave model was introduced in 1998 to increase the forecasting skill in the tropics for tropical cyclone development and path forecasting. In 1999, advancements in the spectral resolution occurred, increasing it to 149 with 40 levels. Spectral resolution increased again in the year 2000 to 255. Finally, in 2006, the ensemble advanced spectrally to 399 and vertically to 62 vertical levels (Persson and Grazzini, 2007, pp. 5–6).

8.1.5 Monthly and Seasonal Forecasting System

The fifth component of ECMWF occurred in 1997 when a seasonal forecasting system was developed and run. The seasonal model displays 2 m temperatures (temperatures at 2 meters (6.5 ft) above the surface of the earth), mean sea level pressures, precipitation, and sea surface temperatures. By 2002, an experimental forecast consisting of monthly predictions was started, and is still relied upon today (Persson and Grazzini, 2007, p. 6). The monthly system is run twice a day, and requires permission from the ECMWF in order to retrieve its data.

8.2 Main US Models

The United States was slower to develop its weather models, with its first main model becoming readily available and relied upon in 1987. This model was called the Nested Grid Model (NGM), and lasted through the year 2000. The NGM consisted of two-level grids: one of the hemispheric scale grid and the other of the synoptic scale grid. This was a special grid at the time as it used the hydrostatic equation within its data calculations. The dependent variables used within the NGM were wind velocity components, potential temperature, specific humidify, and surface pressure. This model was run twice a day at 00Z and 12Z, with a forecast time of 48 hours. The NGM had a vertical resolution of 16 levels with five levels below 850 hPa. Although the NGM stopped being used as a main model in 2000, MOS products (described in Chapter 4, Section 4.7) continued to use data from the NGM into 2009 before it was totally discontinued ("NCEP Nested Grid Model Overview," 1998).

From 2000 to 2002, the Aviation Model (AVN) for short-range forecasts was put into use. The Medium-Range Forecast (MRF) was also developed, but for longer-range forecasting instead ("NCEP Nested Grid Model Overview," 1998). The AVN was different than the NGM as it was run four times a day, with a forecast time of about five days. The MRF was only run once a day, with a forecast time of about two weeks. The MRF was only around for a short time, and in 2002 it was discontinued. The AVN was extended to a forecast time similar to that of the MRF, and was then renamed the Global Forecasting System (GFS) (Stunder, 2010).

8.2.1 Global Forecasting System (GFS)

The Global Forecasting System (GFS) quickly became the leading forecasting model for the United States, and became the leading competi-

tor to the ECMWF. Instead of the limited variables of the NGM and AVN models, the GFS has numerous variables involved in its calculations, ranging from surface variables of land and soil – such as temperatures, wind, precipitation, and soil moisture – all the way to the highest of variables in the stratosphere, such as ozone concentrations. Just like the ECMWF, the GFS covers the entire globe. A major difference to the ECMWF is that it is run four times a day. The GFS has a base resolution of 28 km (17 miles), and currently has a forecast time of 16 days. This resolution is not constant throughout the forecast cycle, which causes some errors and biases, which are discussed later in the chapter. Instead, the resolution changes to 70 km (43 miles) between the one- and two-week periods of the model run. The GFS is composed of four different models all working together to provide an accurate depiction of the weather across the world. These models are the atmospheric model, the ocean model, the land and soil model, and the sea ice model. The GFS is constantly improving and evolving, trying to keep up with newer technologies and computer powers ("Global Forecast System (GFS)," 2015). The GFS has just recently undergone a major upgrade in 2015 and 2016. More information on this upgrade can be found in Chapter 14, Section 14.4.3.

8.2.2 Global Ensemble Forecast System (GEFS)

Just as the ECMWF has an ensemble system, the GFS has its own as well, the Global Ensemble Forecasting System (GEFS). The GEFS consists of 21 separate forecasts to represent and address the uncertainty of the forecasting over a 16-day period and is run four times a day. The GEFS is similar to the ensemble system of the ECMWF in its importance in forecasting and understanding the fluidity and chaos of the atmosphere. ("Global Ensemble Forecast System (GEFS)," 2015). Ensembles play a huge role in forecasting as they allow for many possibilities to be represented in a forecast. While looking at an ensemble mean alone would be an inaccurate tool by itself, it is an extremely valuable complement to the main models and data-accessible to the forecaster. While the ECMWF and GFS may be the best of the main models, they may miss some variable that the ensemble may pick up. The main GFS and ECMWF runs are used in their respective ensemble system as one of the sets of forecast runs. The others used inside the ensemble each have small differences that may or may not play out in reality, but these variables need to be accounted for as it shows how they could potentially alter a forecast. Using the ensemble can show trends of cyclone movement or strength, and can also help show trends in the extended forecast regarding temperatures, or dry or wet spells over certain regions.

8.2.3 North American Model (NAM)

The other main model used by the National Weather Service and the National Centers for Environmental Prediction (NCEP) is the North American Model (NAM). The NAM was first developed in 2006. The NAM is run four times a day and has a forecast period of 84 hours. The value of the NAM is its resolution, which sits at 12 km (7.5 miles). This high resolution allows for much greater detail in forecasting data, and accounts for variables that may be missed in the lower resolution models ("NAM/HIRESW," 2015). This model also includes many parameters, such as temperature, precipitation, and even lightning. The NAM differs from the GFS when it comes to grid size as well, as it only encompasses the North American region, allowing for the high-resolution power, but limiting the areas in which it covers. The high-resolution aspect of the NAM allows for a greater value in extreme weather forecasting and a more accurate depiction of weather structures and behavior. The NAM's ensemble model is called the Short-Range Ensemble Forecast (SREF). This ensemble has the same purpose as the others discussed, and is run four times a day ("North American Mesoscale Forecast System (NAM)," 2015).

8.3 Specific Use Forecasting Models

There are many other computer models accessible and run all over the globe. Some are used for forecasting, others were developed for research purposes, and others are used for more specific needs. Some of these specific models include uses for hydrological needs, road surface needs, and even ocean features. The hydrological model is used to see if heavy rains will produce flooding in the near term or long term. This is a very valuable tool when it comes to monsoonal patterns, or during flash flooding events. This model can be used regarding areas on or along flooding-prone locations, such as rivers, lakes, creeks, seas, or oceans. The road surface model is used to help with temperature at the surface rather than in the air. Tarmac temperatures differ much more than the temperature directly above it in the air, which is often what is used inside the normal forecast models by meteorologists. Traffic as well as the specific heat capacity of the tar or concrete used on roadways has a huge effect on surface temperature, and these models help reflect those data more accurately. These models can predict the road surface temperature during winter storms, which more accurately can show if liquid will freeze on contact, or if snow will melt on contact. Some countries have models linked directly to the road surfaces with sensors installed, relaying information back to be used directly in the models. Models used involving ocean and large water regions consist of swell and wave data. This is of great significance to seaside ports, offshore structures and industries, and shipping (Inness and Dorling, 2013, pp. 145–148).

8.4 Weather Research and Forecasting Model (WRF)

In regards to research, a main model used is the Weather Research and Forecasting model (WRF). The WRF is a next-generation mesoscale

NWP model designed for use in atmospheric research as well as in operational forecasting. The scales of the WRF vary with what is being analyzed and modeled, with scales as small as tens of meters to as large as thousands of kilometers. The WRF simulates the atmosphere by using observed real-time data. The WRF was first introduced in the late 1990s, and is used not only by NCEP and the United States but also in 150 countries across the world ("The Weather Research & Forecasting Model," n.d.).

8.5 Global Data-Processing and Forecasting System (GDPFS)

With the never-ending technological advancements involving computer power, weather model capacity and resolution have greatly increased and continue to do so. Computer models have developed to have a longer lead-time for impending weather events, a large variety and range in relevant applications, and much higher resolutions and accuracy when it comes to forecasting. With all of these advancements it is important to be able to share and collaborate on the data and research produced. The Global Data-Processing and Forecasting System (GDPFS) enables World Meteorological Organization (WMO) members to make use of all these data (Figure 8.2). There is a worldwide network of operational systems under the GDPFS with its main centers called World Meteorological Centers (WMCs). There are three WMC locations: Melbourne, Australia; Moscow, Russia; and Washington DC, in the United States. Along with the main centers, there are 13 centers with operational global NWP capability. These locations are in Australia (BoM), China (CMA), Canada (CMC), Brazil (CPTEC), Germany (DWD), Europe (ECMWF), India (IMD/NCMRWF), Japan (JMA), the Republic of Korea (KMA), France (MF), the United States (NOAA/NCEP), Russia (ROSHIDROMET), and the United Kingdom (UKMO). Appendix Figure A shows each center's NWP system, along with each system's

domain, horizontal resolution, forecast time, and vertical levels, all as of 2013 ("Anticipated Advances in Numerical Weather Prediction (NWP), and the Growing Technology Gap in Weather Forecasting," 2013, pp. 20–23). For a full chart on the most popular weather models worldwide as of 2013 see Appendix Figure A.

8.6 Main Model Biases and Various Characteristics

Although each technological and computer model advancement has tremendously increased the accuracy of meteorological and atmospheric forecasts and trend predictions, there are many common areas of biases and strengths/weakness within each model. Even the higher-resolution models are not without faults, as they tend to miss parts of the world due to their high spatial resolutions. Some terrain, such as mountains or steep gradients, is often missed by these high-resolution models, and the environmental variability that occurs at these locations oftentimes results in lost data (Humans *et al.*, 2005, p. 1). Meteorologists must be aware of this issue when using higher-resolution models, and must also be aware of the other common biases and strengths in the other more commonly used models.

8.6.1 UKMET Characteristics

The UKMET is a very common model used in forecasting across the globe. Its main positive attribute is that it commonly does a better job than the GFS in regards to forecasting the phasing of systems in the northern and southern jet streams ("Subjective List of Model Performance Characteristics," 2013). Some of its more common biases include important aspects that a meteorologist has to be aware of. The UKMET tends to have an eastward bias when developing and progressing upper-level lows, especially over Canada. This model tends to flatten out upper-surface features and patterns over the western Pacific, develops upper

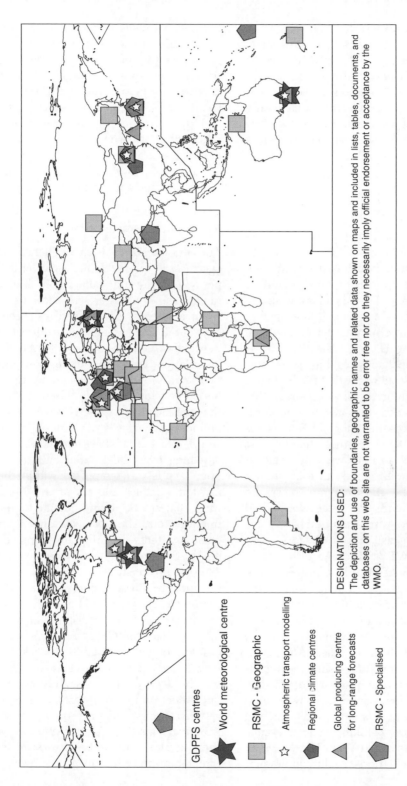

Figure 8.2 The map shows the layout of the different centers in the World Meteorological Organization (WMO). There are three World Meteorological Centers, designated by the larger red star symbol. As seen on this map, locations across the entire world play a major part in the success of the WMO. Note that the figure does not include RCCs in Niger and Brazil. *Source:* Reproduced with permission of WMO. (*See color plate section for the color representation of this figure.*)

systems to far south over the Gulf of Mexico and northwest Caribbean, and keeps the westerlies to far south at times. The UKMET breaks down amplified longwave patterns too quickly as well. Two other common biases of the UKMET involve important features that greatly influence a forecast, as it tends to lower surface pressures too fast as well as having issues representing shallow cold air pools ("Model Biases," 2000).

8.6.2 ECMWF Characteristics

The ECMWF is not immune to its own biases and tendencies either. The ECMWF happens to be very strong when it comes to predicting mid- and upper-troposphere heights during colder parts of the year. This model is also strong at predicting planetary-scale amplitudes such as the Pacific and North Atlantic teleconnection, or PNA. Unlike the UKMET, the ECMWF is good at forecasting shallow cold air situations, and a very valuable trait of the ECMWF is that is has a very small distance error when forecasting springtime lows over the four- and five-day forecast periods. One of the more blatant issues regarding the ECMWF is that this forecast model is only run twice a day, while other global models, such as the GFS, are run four times a day. Some of the other more common biases are that the ECMWF tends to overdevelop closed lows due to its higher resolution, and has a common westward forecast bias of closed lows in general. The ECMWF tends to move shortwave features through zonal patterns at a slower pace ("Model Biases," 2000). Another forecasting error includes misrepresentations in latent heat release. This is important, especially when it comes to cyclonic development where warm and moist air is involved. Another issue is that radiative fluxes are often underestimated, resulting in errors in winter temperature forecasting. The ECMWF has a long-lasting issue with cloud process under thermally stable conditions, and often underestimates blocking pattern events, resulting in a more zonal flow. The ECMWF tends to shift circulations slightly poleward.

It underestimates cyclonic development over the western Atlantic and the blocking frequency during the summer in Europe. The Azores subtropical high pressure is slightly toward the northeast in the summer, and over the northern Pacific the subtropical jet is slightly weakened and the polar jet slightly strengthened. One other common issue with the ECMWF is that during the summer the Inter Tropical Convergence Zone (ITCZ) is slightly too north over Africa (Persson and Grazzini, 2007, p. 41).

8.6.3 NAM Characteristics

The GFS and NAM also have their strengths and weaknesses. The NAM has a shorter forecasting period, having no value when looking long range. The NAM starts to show its strengths when a forecast period reaches three to four days out. It also has a limited grid region, meaning it does not cover the globe, creating rigid boundaries. These boundaries create initialization errors and inaccuracies as the forecast period gets longer within the model because of the guesswork at what is beyond the model boundaries. While other models can view the world, the NAM views only North America. Anything over the Pacific that is outside the boundary is not quantified within the NAM. Certain weather features or atmospheric features that are occurring outside the boundary might be too strong, too weak, not as amplified, or nonexistent when they finally move into the grid as the forecast period moves along in time. This can obviously cause dramatic issues within the model. When it comes to warm weather mesoscale convective systems, the NAM tends to have a slightly lower distance error for storm location than the GFS. When comparing forecast trends with the GFS, the NAM tends to have a north bias, but no east–west bias, while the GFS seems to have a slightly less north bias, but a more pronounced east bias, moving storm systems through the grid too quickly (Yost and Schumacher, 2012). The NAM in comparison to the GFS also tends to be historically less accurate in the winter season, greatly overestimating precipitation totals, possibly due to added and incorrect convective feedback ("Subjective List of Model Performance Characteristics," 2013).

8.6.4 GFS Characteristics

The pre 2015 upgraded GFS tends to perform relatively well when it comes to pattern changes, seeing colder or warmer air changes further out in time. Having four model runs each day is also of huge value when it comes to the GFS. These model runs have issues within the setup, however, as the resolution changes after the 10-day period. This resolution change in the GFS creates timing issues and unrealistic solutions at times. Because of this, the latter parts of the model run are usually not taken as specifics, but rather as guides to what may occur, or as trends. From the time the resolution changes through the end of the period most of the analysis that occurs is about patterns that may occur or start to show themselves. Arctic air intrusions or monsoonal patterns may start to show face. From about a week out to the change in resolution, more distinct air mass regions can be seen, with a better feel on jet stream locations and amplifications of troughs and ridges. From current time to about a week out, that is when actual weather features can be more accurately forecast, with timing and precipitation totals, temperatures, and wind speeds and directions starting to become more accurate and dependable. The GFS's more prevalent biases relates to precipitation totals. In the warm season, the convective parameterization scheme does not fully represent convective feedback accurately. The GFS also tends to have a dry bias north of areas where two inches of total precipitation is produced over a six hour time period. These two issues seem to be related to each other. The NAM however, tends to perform better than the GFS in this area when it is in the warmer seasons. The GFS also commonly over does the aerial coverage of mass fields of total precipitation of zero to one tenth of an inch, likely due to its resolution issues. Another resolution issue leads to the GFS becoming too

ambitious with its strength and speed of systems crossing over the Sierra Mountains after the 36-hour forecasting period in the cool seasons. This is likely due to the combination of model resolution issues and the topography of the region. It is also slightly ambitious when it comes to the magnitude of high amplitude patterns, and when it comes to phasing of the northern and southern jet streams. The GFS tends to phase the jets too fast, which is likely the result of too few weather observational data in the weather system's origination point in the forecast cycle ("Subjective List of Model Performance Characteristics," 2013).

8.6.5 Ensemble Characteristics

As previously mentioned, the ensembles have large value when it comes to forecasting. Getting the average of many model runs each with a slightly different set of variables attached to them allows for the limits of uncertainty to be represented and quantified. This too, however, does not come without any biases attached to it. The ensembles have a mean for the maximum and minimum of many different forecasting components, such as temperatures, wind speed and direction, pressures, precipitation, etc. This mean is not always the best representation of all the data, as the mean can sometimes be washed out, or subdued, when compared to the actual weather verification. This occurs because the averages tend to flatten out the amplitude of ridges and troughs, creating a misrepresentation for what will most likely occur ("Subjective List of Model Performance Characteristics," 2013). Specifically, the GEFS ensemble mean at week one and week two tends to also have large fluctuations on a day-to-day basis when it comes to mean precipitation and mean temperature. The GEFS does happen to have a clearer seasonal cycle with better skill scores in the winter when it comes to mean temperature and precipitation. The errors with these two specific components tend to be more regional and time-of-year-dependent (Fan and Van Den Dool, 2011, pp. 369–370).

Chapter Summary

From the 1980s through the 1990s and the early twenty-first century, computer models became what we see them as today, as a vital tool for the meteorologist and forecaster. This tool is not the end of all tools, though, as many biases and weaknesses pop up within each model that is used. A forecaster must know what the weaknesses and biases are, as well as which model is stronger at certain components than others. There are numerous models out there to aid meteorologists and forecasters, many more than are discussed in this chapter. Organizations such as the WMO and the GDPFS allow for the sharing of data and research from most of the computer models out there, helping to further the field and unite ideas and information. Whether a need arises for more specific weather needs, such as road temperature, whether there is an extreme weather event in a localized location where high-resolution models are needed, or whether you are looking for a trend in predicted weather patterns or monthly forecasts, there are computer models that are easily accessible and trusted to help acquire the information needed.

References

Book References

"Anticipated Advances in Numerical Weather Prediction (NWP), and the Growing Technology Gap in Weather Forecasting." (2013) *WMO*, Submitted by WMO, with contributions from the CBS/DPFS chairperson, United Kingdom.

ECMWF (2012) *The European Centre for Medium-Range Weather Forecasts: European co-operation at its best*, ECMWF, Reading.

Inness, P. M. and Dorling, S. (2013) *Operational Weather Forecasting*, John Wiley & Sons, Ltd., Chichester.

Persson, A. and Grazzini, F. (2007) *User Guide to ECMWF Forecast Products*, Version 4.0, 14 March 2007, Meteorological Bulletin M3.2, ECMWF, United Kingdom.

Journal/Report References

Fan, Y. and Van Den Dool, H. (2011) Bias Correction and Forecast Skill of NCEP GFS Ensemble Week-1 and Week-2 Precipitation, 2-m Surface Air Temperature, and Soil Moisture Forecasts, *Weather and Forecasting*, 26 (3), NOAA/NCEP/CPC, Camp Springs, Maryland, pp. 369–370.

Humans, R.J., Cameron, S.E., Parra, J.L., *et al.* (2005) Very high resolution interpolated climate surfaces for global land areas, *International Journal of Climatology*, 24: 1965–1978.

Yost, C.M. and Schumacher, R. (2012) Do the Global Forecast System (GFS) and the North American Mesoscale (NAM) Models Have Displacement Biases in Their Mesoscale Convective System Forecasts?, *Departments of Atmospheric Sciences*, Texas A&M University and Colorado State University, Fort Collins, Co.

Website References

"Global Ensemble Forecast System (GEFS)." (2015) *Datasets*, National Centers of Environmental Information, NOAA, https://www.ncdc.noaa.gov/data-access/model-data/model-datasets/global-ensemble-forecast-system-gefs, accessed 11 September 2015.

"Global Forecast System (GFS)." (2015) *Datasets*, National Centers of Environmental Information, NOAA, https://www.ncdc.noaa.gov/data-access/model-data/model-datasets/global-forcast-system-gfs, accessed 11 September 2015.

"Model Biases." (2000) *Weather Prediction Center*, NWS and NCEP, http://www.wpc.ncep.noaa.gov/mdlbias/biastext.html, accessed 11 September 2015.

"NAM/HIRESW." (2015) *Environmental Modeling Center*, NWS and NOAA, http://www.emc.ncep.noaa.gov/index.php?branch=NAM, accessed 11 September 2015.

"NCEP Nested Grid Model Overview." (1998) *Numerical Weather Prediction Links*,

National Weather Service, http://www.srh.noaa.gov/ssd/nwpmodel/html/ngmover.htm, accessed 9 September 2015.

"North American Mesoscale Forecast System (NAM)." (2015) *Datasets*, National Centers of Environmental Information, NOAA, https://www.ncdc.noaa.gov/data-access/model-data/model-datasets/north-american-mesoscale-forecast-system-nam, accessed 11 September 2015.

Stunder, B. (2010) *What is the Difference Between the AVN, MRF, and GFS Models?*, Air Resources Laboratory, NOAA, https://www.ready.noaa.gov/faq_md15.php, accessed 11 September 2015.

"Subjective List of Model Performance Characteristics." (2013) *Weather Prediction Center*, NWS and NOAA, http://www.wpc.ncep.noaa.gov/mdlbias/biastext.shtml, accessed 11 September 2015.

"The Weather Research & Forecasting Model." (n.d.) *WRF*, UCAR, http://www.wrf-model.org/index.php, accessed 11 September 2015.

Figure References – In Order of Appearance

"Member States." (n.d.) *ECMWF*, http://www.ecmwf.int/en/about/who-we-are/member-states, accessed 20 July 2016.

"Anticipated Advances in Numerical Weather Prediction (NWP), and the Growing Technology Gap in Weather Forecasting." (2013) *WMO*, Submitted by WMO, with contributions from the CBS/DPFS chairperson, United Kingdom, p. 4.

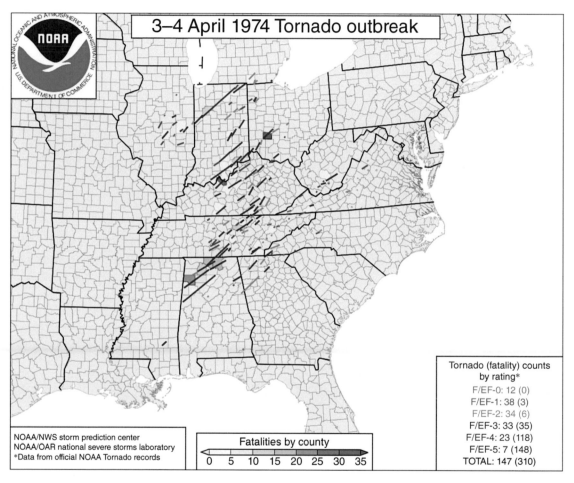

Figure 7.2 The map shows the locations and distances of the tornado outbreak of 3–4 April 1974 in the United States. The table in the bottom right corner shows the total number of each category of tornado, along with the fatalities that occurred in each category tornado. In total, 147 tornadoes occurred, resulting in 310 deaths. Tornadoes spanned from Alabama northward into Michigan, and as far east as central Virginia. *Source*: Courtesy of NOAA.

The Evolution of Meteorology: A Look into the Past, Present, and Future of Weather Forecasting, First Edition.
Kevin Anthony Teague and Nicole Gallicchio.
© 2017 John Wiley & Sons Ltd. Published 2017 by John Wiley & Sons Ltd.

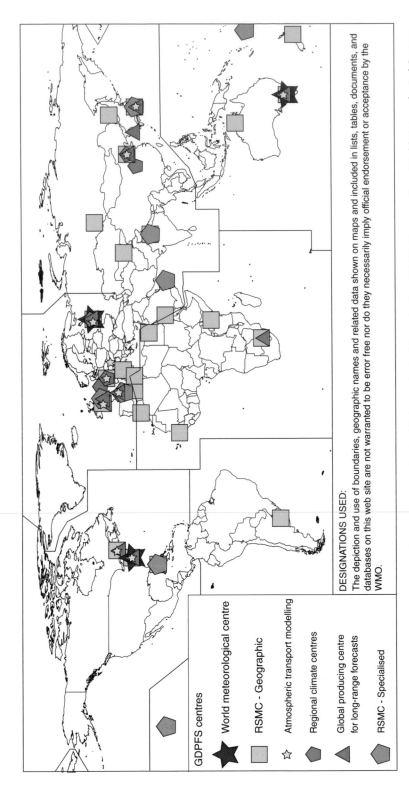

Figure 8.2 The map shows the layout of the different centers in the World Meteorological Organization (WMO). There are three World Meteorological Centers, designated by the larger red star symbol. As seen on this map, locations across the entire world play a major part in the success of the WMO. Note that the figure does not include RCCs in Niger and Brazil. *Source:* Reproduced with permission of WMO.

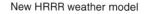

Existing weather model   New HRRR weather model

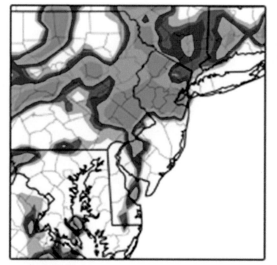 

Figure 14.6 In 2014, the NWS started to implement a high-resolution weather model into its toolkit for weather forecasters and researchers. The images are of the same storm. The difference between the two models is extremely noticeable, and shows just how valuable high resolution is. As weather agencies develop more powerful computers, the better they are then equipped to run high-resolution models, greatly increasing accuracy in forecasts, even during localized thunderstorm outbreaks such as the case displayed here. *Source*: Courtesy of NOAA Photo Library.

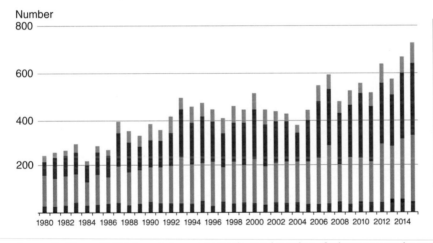

NatCatSERVICE

Loss events worldwide 1980–2015
Number of relevant events by peril

Geophysical events
(Earthquake, tsunami, volcanic activity)

Meteorological events
(Tropical storm, extratropical storm, convective storm, local storm)

Hydrological events
(Flood, mass movement)

Climatological events
(Extreme temperature, drought, forest fire)

Accounted events have caused at least one fatality and/or produced normalized losses ≥ US$ 100k, 300k, 1m, or 3m (depending on the assigned World Bank income group of the affected country).

Appendix Figure C This image demonstrates the total number of relevant events by peril from 1980 to 2015. The total number of events causing death has increased steadily, especially over the past decade. Meteorological and hydrological events are among the greatest contributors to loss of life. *Source*: Reproduced with permission of Munich Re (NatCatSERVICE).

Figure 9.5 The difference in technology from 1970 to today is extremely dramatic. One example of how far we have come since 1970 is when looking at satellite images and quality. To the left is an image from the 1970s from the TIROS 1 satellite of the Gulf of St. Lawrence and the St. Lawrence River. The image on the right was taken in 2013 from the Suomi NPP satellite of New England. *Source*: Courtesy of NOAA Photo Library.

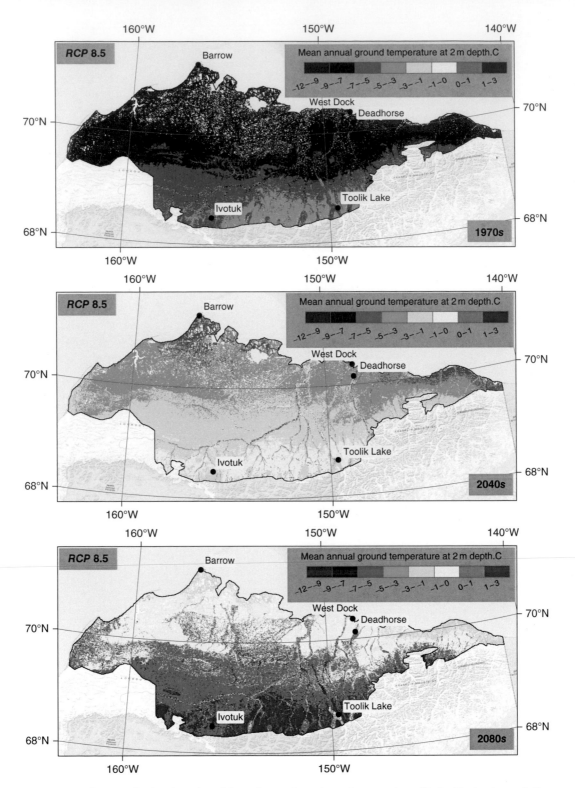

Figure 11.2 The spatially distributed modeling of permafrost throughout northern Alaska. The top image is the temperatures present in the 1970s. The bottom image is an estimate of the predicted temperature in the 2080s. The most drastic observation one can take from this comparison is the all-round general warming trend in all areas. Looking at the future predictions, temperatures are estimated to be above freezing for nearly all of northern Alaska, with the coldest locations to the far north at only around −1 to −3 °C. This is compared to the 1970s, when the warmest locations were only as high as −3 to −5 °C and as cold as −12 °C to the far north. *Source*: Courtesy of Dr. Dmitry Nicolsky.

Loss events worldwide 1980–2015
Number of severe catastrophes by peril

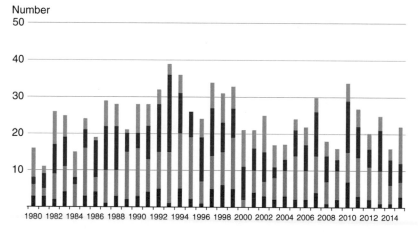

Geophysical events
(Earthquake, tsunami,
volcanic activity)

Meteorological events
(Tropical storm,
extratropical storm,
convective storm,
local storm)

Hydrological events
(Flood, mass movement)

Climatological events
(Extreme temperature,
drought, forest fire)

Accounted events have caused ≥1,000
fatalities and/or produced normalized
losses ≥ US$ 100m, 300m, 1bn, or 3bn
(depending on the assigned World Bank
income group of the affected country).

Appendix Figure D While events causing at least one loss of life has grown steadily, severe catastrophes (events causing 1000 or more deaths) have decreased recently, especially when compared to the 1990s. This may be due to increased technology and forecasting power leading to better warnings, better forecasting and predictions, and better relief efforts and planning. This image also shows how most recent events causing severe loss of life are mostly attributed to climatological events, such as extreme temperatures, droughts, or forest fires. *Source*: Reproduced with permission of Munich Re (NatCatSERVICE).

NatCatSERVICE

Natural loss events worldwide 2015

Geographical overview

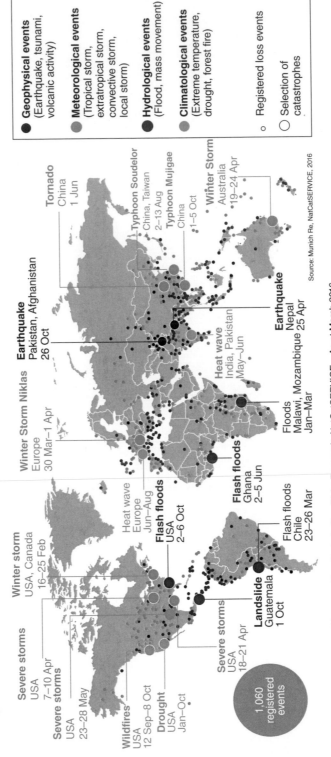

Source: Munich Re, NatCatSERVICE, 2016

@2016 Münchener Rückversicherungs-Gesellschaft, Geo Risks Research, NatCatSERVICE – As at March 2016

Appendix Figure G This image shows the natural loss events worldwide in 2015, including 1060 registered events, either geophysical, meteorological, hydrological, or climatological in nature. This image shows that no area is safe to extreme weather or conditions. Some of the more devastating events are highlighted with dates of occurrence.

9

Advancements within the Nationalized Governmental Weather Departments

Kevin Anthony Teague

Not only did the computer models expand drastically from the 1980s through the turn of the century but also the departments and agencies that were home to those computer models did as well. With an expansion and technology boom in regards to the sciences surrounding satellites and radar, nationalized weather departments became the go-to hubs of information, research, and consulting in areas of weather forecasting and analysis for country leaders, policymakers, and various other departments and agencies. The advancements that have occurred from what was prevalent 50 years prior all the way to today are almost incomprehensible. Fifty years ago, weather departments were first witnessing the importance and significance that radar, satellites, and computers had on forecasting. Today, we are using high-resolution computer models, dual Doppler radars, and multiple types of satellites accessible at a click of a button on a home computer. So much has changed in just 50 years, and the departments responsible for those changes continue to adapt to new technologies and needs every day.

9.1 The National Weather Service

From the 1980s through the first decade of the twenty-first century, the greatest advancements within the entire government of the United States was arguably within the National Weather Service (NWS). Throughout the United States, national significance was placed on the NWS to forecast for many of the nation's most important events and natural disasters. Starting in 1984, the NWS was in charge of meteorological forecasting and consulting in regards to the Olympic Games that were held in Los Angeles. The NWS was then also used for the Olympic Games in 1996 in Atlanta, and again in 2002 in Salt Lake City. Sporting events were not the only national responsibility that the NWS was faced with. In 1986, the US *Voyager* accomplished the first ever non-stop non-refueled flight around the world, aided by former and active meteorologists from the NWS. When it came to tropical implications, the NWS stepped up to the task, with strong examples dating back to Hurricane Hugo in 1989, saving countless lives and property damage by issuing accurate and dependable forecasts from the Miami Hurricane Center. In 2005, the NWS and the National Hurricane Center had their most historic year, naming 28 storms (at least 39 mph or 35 kts), 14 of which were hurricanes (74 mph or 65 kts), with seven of them categorized as major (category 3 or greater on the Saffir–Simpson hurricane wind scale, 111+ mph or 96+ kts). Of those major hurricanes, three were category 5 of at least 157 mph or 137 kts: Katrina, Rita, and Wilma. Altogether that year, four major hurricanes and three tropical storms made landfall in the

The Evolution of Meteorology: A Look into the Past, Present, and Future of Weather Forecasting, First Edition.
Kevin Anthony Teague and Nicole Gallicchio.
© 2017 John Wiley & Sons Ltd. Published 2017 by John Wiley & Sons Ltd.

United States, putting a huge significance and value on the work done by the NWS ("History of the National Weather Service," n.d.).

The NWS serves as a major contributor to the safety and well-being of the main population. It got to that point by undergoing a great modification throughout the entire agency starting in 1989, when the NWS set forth a very elaborate national plan for the modernization and restructuring of the department. The main goal of this $4.5 billion overhaul was to change the operational methods used throughout the agency, to improve in all aspects in order to better save lives and livelihoods, and to develop and implement five different technologies. These technologies included how observations were to be gathered, to update the national radar system, to launch new satellites, to advance the computer systems within the individual centers, and to update the programs used within the newer computers ("History of the National Weather Service," n.d.).

9.1.1 Automated Surface Observing System

Prior to this plan, weather observations were done manually. Humans had to report what was occurring or had occurred, leaving significant room for human error and limiting the accessibility of observations when observations were needed at specific times for specific weather events. This plan aimed to make weather stations automated, calling them the Automated Surface Observing System (ASOS). This system made weather observations more clear, more readily available, more accurate, and, most importantly, took the human element out of observations, leveling the accuracy and continuity between all of the ASOS stations.

9.1.2 Next Generation Weather Radar

The Next Generation Weather Radar (NEXRAD) was implemented to greatly overhaul and update the radar system throughout the entire country (Figure 9.1). NEXRAD was one of the more significant advancements. The main objective of this technological advancement for the NWS

was to develop a radar system that was connected nationally. Each radar was built with advanced Doppler radar improvements. This was designed to increase lead-time in severe weather forecasting and nowcasting, in events with tornadoes, flash flooding, hail, hurricanes, and more. Full-scale production of 165 NEXRAD units and over 300 display substations began in 1997 ("History of the National Weather Service," n.d.).

9.1.3 Satellites

The NWS placed a lot of significance on the power and meteorological importance of weather satellites during this time. Satellites were able to see the world's weather from a unique perspective, allowing for longer-term forecasting and a better knowledge of storm system behavior. With each satellite that was launched, newer technology and newer data were able to be used and retrieved, greatly enhancing the forecasts delivered by NWS meteorologists. See Section 9.5 for more on satellites during this time.

9.1.4 Advanced Computer Systems

Advancing the computer systems was a no-brainer for the NWS during this modernization plan. The NWS was greatly seen as being behind the times, especially when compared to the Met Office. Increasing computer power was at the forefront of almost all these advancements, as any new radar data or satellite data that were retrieved needed the stronger-powered computers to house and store them. During this time, weather models were expanding, needing the speed and power as well in order to be fully operational. In 1990, the supercomputer Cray Y-MP8 was installed, bringing with it high-resolution technology and much more sophisticated numerical weather prediction (NWP) capabilities. By 1995, the NWS launched its Internet Service Interactive Weather Information Network (IWIN). The advancements didn't stop there, continuing again in 2007 when the newest weather and climate supercomputer became available.

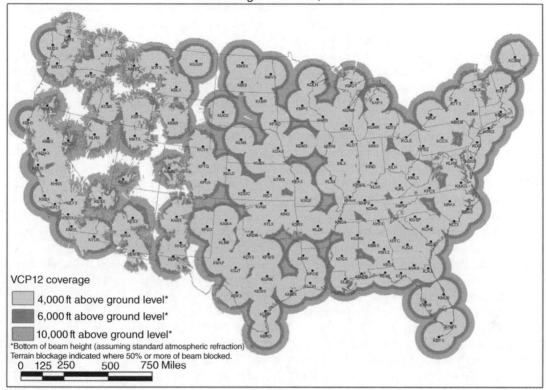

NEXRAD coverage below 10,000 feet AGL

VCP12 coverage

- 4,000 ft above ground level*
- 6,000 ft above ground level*
- 10,000 ft above ground level*

*Bottom of beam height (assuming standard atmospheric refraction)
Terrain blockage indicated where 50% or more of beam blocked.

0   125 250   500   750 Miles

Figure 9.1 The map displays the NEXRAD weather radar system across the contiguous United States. While there are some gaps in coverage across the Rocky Mountain region, most of the country is covered by weather radar, including much of the eastern half of the country with many locations overlapping with multiple radar coverage. *Source*: Courtesy of NOAA Photo Library and NWS.

This IBM machine produced a 320% increase in power and calculation speed, increasing production speed to 14 trillion calculations per second, allowing for the implementation of over 240 million global weather observations a day. By 2009, the NWS again increased its computer power by installing the next generation of IBM supercomputers at a cost of $180 million, but with the benefits of the production speed of 69.7 trillion calculations per second ("History of the National Weather Service," n.d.).

9.1.5   Advanced Weather Interactive Processing System

The fifth technological improvement for the modernization plan was with the weather program and communication system used within the offices of the NWS, called the Advanced Weather Interactive Processing System (AWIPS). The new AWIPS system was set up to allow offices to better communicate with one another, leading to a greater understanding of storms that were coming. Information that may not be available at one office can be shared from another, adding to the continuity between offices and the NWS as a whole. The AWIPS system also allowed for easier distribution of data and offered field forecasters access to the data that was available within the office. In 2000, AWIPS was installed in 152 sites across the country. This is considered the final stage of the modernization and restructuring of the NWS ("History of the National Weather Service," n.d.).

AWIPS in general is an interactive computer system with the capabilities of integrating all the major data elements into one centralized location for the first time ever, including meteorological, hydrological, satellite, and radar data. Prior to this modernization, a meteorologist at the NWS had to view all weather data separate from each other, which necessitated several different machines to be able to look at images or data side by side. The workstation for a meteorologist at the NWS with AWIPS consists of three graphic monitors, each able to produce up to 15 different windows of weather information and data, and one text workstation. Radar and satellite data can be overlaid and observed simultaneously, with data such as lightning strikes, upper-air dynamics, surface observations, balloon data, and model guidance's all looped, overlaid, and zoomed in on each other ("AWIPS Advanced Weather Interactive Processing System," n.d.) (Figure 9.2).

The text work prior to the modernization was very crude, black and white, with a maximum of three products that could be overlaid. The ability to loop text was not only crude but extremely slow, and the word processing in general was extremely time consuming and difficult, greatly limiting the full capabilities of the meteorologist's ability to relay a forecast in a timely manner. The modernization led to a much better text station, called the Interactive Forecast Preparation System (IFPS) and the Graphical Forecast Editor (GFE). In 2002, this advancement was implemented, creating regular text-based products, digital graphics, and forecasts. Prior to this advancement, meteorologists had to manually enter data, but with this advancement, similarly to the ASOS advancement, the human element was removed, with forecasts imported from computer models and weather data into a GFE. The meteorologist could then use their time and knowledge more efficiently, adjusting temperatures, precipitation probabilities and amounts, wind elements, and other weather elements within the forecast on the interactive editor. This editor is another example of the advancing idea of continuity and nationalized network of data, as it created easily accessible and detailed weather mapping, text, and forecasts across the country all in one National Digital Forecast Database (NDFD) ("IFPS/GFE Interactive Forecast Preparation System Graphical Forecast Editor," n.d.).

(a)

(b)

Figure 9.2(a) The Advanced Weather Interactive Processing System (AWIPS), used across the NWS in the United States. The three graphic monitors to the left can display up to 15 different windows, including radar, satellite, weather balloon data, NWP products, observed conditions, and lightning data. The text station used to disseminate forecasts and warnings, along with allowing forecasters to look at weather discussions and text observations. (b) The computer rack needed to support the AWIPS system. *Source*: Courtesy of the NWS and the Upton, New York Office.

(c)

Figure 9.2(c) The Interactive Forecast Preparation System Graphical Forecast Editor (IFPS/GFE) takes away the tedious task of manually entering data and information to create a weather graphic. Instead, the IFPS allows forecasters to import data into the GFE and then manipulate the images for each specific type of forecast. Such forecasts include temperatures, precipitation probabilities and amounts, wind speed and direction, and wave heights. *Source*: Courtesy of the NWS and the Upton, New York Office.

9.1.6 Further Advancements

Even after the modernization plan was complete, the NWS had to again revamp its radar network to keep up with advancements in technologies. In 2011, the NWS implemented its next phase of future advancements and areas of focus, calling it the need to be a "weather-ready nation." This movement was designed to make America a safer nation from meteorological, atmospheric, and climatological influences. The greater technological advancements resulted in a new need for space weather forecasting as solar storms created the potential for great impact on satellites, computers, the Internet, cellphones, and more. An increased population also leads to problems and greater risks of disaster, especially when connected to severe weather, tornado outbreaks, heatwaves, flooding, tropical disturbances, and winter storms. With these new issues combined with further solving the old and current issues in forecasting and weather knowledge, the NWS continually tries to improve its operations. This can be seen by its ranking in the top 15% of all US federal agencies in customer satisfaction ("History of the National Weather Service," n.d.).

The NWS under its Weather-Ready Nation initiative began to upgrade its network of Doppler radars. The new advancement with this set of radar technology was dual-polarization (or dual-pol). Dual-pol radar greatly increases the accuracy of rain forecasts, hail detection, precipitation type detection, and more. More on dual-pol technology can be found in Section 9.4.

9.2 The Met Office

The Met Office also saw its importance explode during this period of time. This was seen in 1984 when the Met Office was named a World Area Forecasting Center (WAFC), for civil aviation. At this time, the only other WAFC was in the United States. Further importance was placed on advancing the Met Office after the Great Storm of 1987 (see Chapter 7, Section 7.1).

After this event, the Met Office developed the NSWWS, or the National Severe Weather Warning Service, for all the United Kingdom. This service was created to provide warnings to the public and to protect life and property. By 1990, the Met Office opened the Hadley Center, which was a center dedicated for the research of the climate. This center is widely seen as the pioneer of research in climate change, often advising the governments of the United Kingdom, the United States, and many others on its research and observed trends and forecasts. The Hadley Center is often at the front of global policy changes when it comes to climate issues ("Overview of the Met Office," 2012).

These examples of the value placed on the Met Office do not demonstrate the full extent of the Met Office's influence throughout the world. By 2012, the Met Office implemented supercomputers that produce 3000 forecasts and briefings a day, delivering them to governmental agencies, businesses, policymakers, the armed forces, the public, and many others. In 2012, the Met Office had 1700 employees in 60 locations around the world. It is recognized "as one of the world's most accurate forecasters, using more than 10 million weather observations a day, an advanced atmospheric model and a high performance supercomputer to create 3,000 tailored forecasts and briefings a day" ("Overview of the Met Office," 2012). This worldwide grasp "provides vital services, advancing global understanding through research, being an important participant in projects and organizations" ("International Role," 2013). The Met Office developed into an agency with a wide skillset attached to it, ranging from working with hospitals by showing the relationship between weather and hospital admissions and medical conditions to general warnings for public safety, and to advising transportation departments and companies on road, air, and sea conditions. The Met Office also advises on electricity, oil, and gas output and usage. The Met Office is a highly influential government department in the United Kingdom, and yet its influence goes beyond government and into the private sector, and beyond the United Kingdom, to Europe and much of the rest of the world (see Chapter 10, Section 10.5.4) ("Overview of the Met Office," 2012).

9.2.1 The Met Office around the World

In 2011, two out of the top five countries in the world that were most impacted by weather-related loss were located in South-East Asia. The Met Office has not turned a blind eye to this fact and instead has put a lot of its resources into this issue. The Met Office aims to reduce economic and social impacts that weather, climate, and other environmental events and hazards have on this region. Many of the more common issues are storm surges from strong tropical systems and typhoons or cyclones, along with flooding, and even tsunamis. The Met Office consults and holds training for meteorological service providers. Its numerical modeling system is open and available to be used by a variety of organizations around the world. The high-resolution forecasting created by the supercomputers and advanced technology within the Met Office allows for forecasts to be relevant to any location in the world, allowing better consulting and advising for regions without those capabilities and technologies to be readily available. Reducing risk and increasing efficiency are two of the Met Office's main goals, but it is also heavily involved in renewable energies, and the decision makers influencing that industry. In many areas around the world, the Met Office issues ocean and water forecasts for offshore and near-shore industries and personnel, reducing the risk of weather-related injuries, costs, and increasing safety and production ("The Met Office in Southeast Asia," 2014).

9.2.2 Public Weather Service

Another main responsibility of the Met Office is the Public Weather Service (PWS). The PWS provides weather forecasts to the public to help with day-to-day activities. The NSWWS is part of the PWS and aids in the safety of the public

("Overview of the Met Office," 2012). The PWS is funded by the UK government not only to produce forecasts to help warn the citizens but also to improve research regarding weather and climate prediction. The PWS makes sure all international commitments are fulfilled and provides public access to historical weather information and data. It is overseen by the Public Weather Service Customer Group ("Public Weather Service," 2015).

9.2.3   The Met College

The Met Office is not only a worldwide source of weather information, consulting, advising, research, and forecasting but also an educational resource as well. Dating all the way back to 1939, the Met Office College began as a training school, training and educating military personnel in weather and how weather might impact military operations. By 1945, the Met College grew and began training students from around the world. By the turn of the century, the Met College began issuing National Vocational Qualifications (NVQs), for forecasters and observers. As of 2015, 275 certificates have been approved by the Met College. These qualifications are recognized nationally as well as by the European Union. Some of the more popular and recent students of the Met College include Prince Harry and the Duke of Cambridge, as they were educated as part of their military training. In 2012, the Met Office began MOCO (Met Office College Online). This allows for virtual learning designed for weather observers, forecasters, and people who want to just simply better their overall weather understanding and knowledge ("History of the Met Office College," 2015).

9.3   Advancements across the World

The United States and the Met Office were not the only weather departments that were growing in importance and advancements during the 1980s through the first decade of the twenty-first century. Countries across the world developed and modernized their own weather departments as weather became a much larger player in trade, travel, and everyday life. Extreme disasters occur all over the world and it is a necessity for countries to develop their own meteorological departments. Meteorological forecasting and research in a specific area allows for the departments to better suit their country's needs and preparations for disastrous weather events.

Many developing countries and developing economies were at risk of weather disasters and hazards. The less developed nations were poor and less resistant and therefore more prone to natural disasters. These types of locations lacked the capacity to provide the proper knowledge, information, and warnings to their citizens, leaving many if not all citizens at risk of impending hazards without any notice. On top of the risk of unknown disaster, the cleanup and recovery is sometimes just as devastating and often unsuccessful without the help of outside stronger nations. A lot of dependence has been placed on outside help (Rogers and Tsirkunov, 2013, p. 17). Some of that outside help comes from organizations such as the World Meteorlogical Organization (WMO) or the National Meteorological and Hydrological Services (NMHS). (See Chapter 10 for more on world organizations.) The NMHS is the "primary operational service provider of observations and predictions to support a country's efforts to mitigate climate hazards and the impact of extreme weather needs" (Rogers and Tsirkunov, 2013, p. 17). The NMHS plays a huge role by providing these less fortunate and developing nations and regions with products and services that would otherwise be unavailable. The WMO assists the NMHS by providing training and weather data and information, and by providing and explaining the NWP models and satellite data as part of the WMO international cooperation agreements. This cooperation is fundamental to the safety of developing nations and their economies, but not all the work can be done by outside help. A lot depends on the nations in need of help helping themselves. These nations have to keep up with their own advancements, through technology, policy changes, and industrial budgets and outlines. The main goal is to level the playing field when it comes to worldwide weather and environmental knowledge to keep all humans safe, but if the less developed nations do not try to help themselves, the gap between the advanced nations and the less advanced nations will only get wider as newer technology and resources leave citizens of the lower economic nations further behind (Rogers and Tsirkunov, 2013, p. 19). The WMO plays a similar role as the NMHS, although in a much greater capacity. The WMO helps modernize the NMHS, focusing its resources in the locations and regions that are attempting to help themselves and showing stability in their own central governments, such as many nations in the Caribbean (Rogers and Tsirkunov, 2013, p. 123). Elsewhere, attempts to help are often stalled or withdrawn completely due to weaknesses in public services, a lack of governmental commitment, low population, lack of available resources to design newer and updated systems, and a lack of trained personnel (Rogers and Tsirkunov, 2013, p. 124).

There are many locations throughout Europe and Asia that are not at the level of some of the more developed nations within these continents. Parts of Europe and Asia have deficiencies in their numerical modeling capacity, partly because of slower advancements in computer technology but also due to sparse and rundown monitoring networks and weak telecoms. Oftentimes, due to some of the previously mentioned reasons, training is out of date, creating gaps in the skills of staff and the trainers themselves (World Bank, 2008, p. XVII). Less-developed nations often have resources and technology that date back decades. This plays a major role in the insufficiencies of these areas in regards to weather forecasting. The technology of the 1980s created a skill score of a two- to three-day forecast equal to the skill score of today's seven-day forecast. Today, forecasters often place a value on 10-day forecasts that have better or equal skill scores to that of the five- to seven-day forecast of the 1980s (World Bank, 2008, p. XVIII). The countries and regions stuck with this older technology are at huge disadvantages when it comes to accurate forecasting. Oftentimes, developing nations state that costs are a main reason for the disadvantages and gaps in their weather departments. This is hotly debated, as many believe that with "weather satellites, supercomputing, and models all being available to low income countries for little more than the incremental cost of connecting to the system" that the benefits these technologies and advancements bring will exceed the costs. (World Bank, 2008, p. XX).

While many nations across the world still require the help of global organizations, many others have grown their own departments during this time. Some of the greater advancements took place in the nations of, but not limited to, South Korea, Japan, South Africa, Australia, China, Russia, and Mexico. These nations emphasize the growing field of meteorology and meteorological knowledge with all of their achievements, advancements, departmental setup, and overall missions.

9.3.1 Korea Meteorological Administration

South Korea saw some of its greatest advancements during this period. Starting in 1979, the Korea Meteorological Administration implemented its computer communication systems. Throughout the next two decades, computer advancements were made, finally leading to the development of a supercomputer by 2000. By 2007, the Korea Meteorological Administration was honored by being made a member of the WMO Executive Council ("Brief History," 2009). South Korea focuses on becoming a leader in promoting public safety and national growth through its advanced meteorological services, placing value on in-time, accurate, and valuable weather information, forecasting, and research ("Mission & Vision," 2009).

9.3.2 Japan Meteorological Agency

Japan, always thought of as a technologically strong nation, further increased the safety of its citizens in 1991 by adding an important seismic intensity meter to its observational data collection. In 1993, the Japan Meteorological Agency revised its Meteorological Services Act (see Chapter 10, Section 10.1) in order to place significance on the quality of its work and research. This act established a certified weather forecasting system ("History," n.d.).

9.3.3 South African Weather Service

The vision statement for the South African Weather Service shows the value that the nation places on its weather department, stating that it is a "weather and climate center of excellence providing innovative solutions to ensure a weather ready region, sustainable development, and economic growth" ("About Us," 2015). The South African Weather Service is a public entity that has worked under the Ministry of Water and Environmental Affairs since 2001. It consists of two main branches: a public good service provider and a paid-for commercial services provider. The South African Weather Service consists of 214 automatic weather stations, 50 weather buoys in both the South Atlantic and South Indian Oceans, 14 weather radars, and 10 upper air-sounding stations ("Overview," 2015).

9.3.4 Australian Government's Bureau of Meteorology

The Australian government's Bureau of Meteorology consists of five major divisions. The first of which is the Hazards, Warnings, and Forecasts Division. This division includes areas of weather forecasting, flood forecasting, hazard predictions, and regional forecasting. It is also the division that houses the Bureau of National Operation Center. The second division is the Observations and Infrastructure Division. This division consists of branches designated to observing strategies and operations, infrastructure management, and regional operations.

The third division is the Environment and Research Division. This division consists of the branches in charge of climate information services, water information services, environmental information services, and research and development. The fourth division includes areas of information technology, system development and maintenance, environmental information management, and digital data delivery, all under the Information Systems and Services Division. The fifth and final division of the Bureau of Meteorology is the Corporate Services Division, where finances and budgets are run, where the branches involved in people management and business development are represented, and where the international branch is stored. The large breakdown of divisions and services within each branch shows the vastness of this one agency, and how vital a role it plays in the lives of all Australian citizens and in the world as well ("Organisational Structure," 2015).

9.3.5 China Meteorological Administration

China realizes the value in a strong meteorological department because 400 million people each year, spanning five different climate zones, are bound to be affected by meteorological hazards. The Chinese use a lot of resources to better advance the China Meteorological Administration in order to help protect the nation and the livelihoods of its citizens. It is estimated that China loses between 1 and 3% of its gross domestic production (GDP) every year because of solely environmental reasons, and the China Meteorological Administration attempts to help limit these loses. The Chinese have developed a strong foothold in exploring wind and solar energy, weather modifications research, lightning detection, space weather monitoring, and atmospheric composition monitoring. The China Meteorological Administration has established an "integrated meteorological observing system incorporating space-based, airborne, and groundbased observations," increasing accuracy of weather forecasts and climate projections ("Message of Admini-strator," 2011). The Chinese are also huge players in advancing the technology and monitoring of the world through their satellite developments.

9.3.6 Hydrometeorological Centre of Russia

The principal task of the Hydrometeoro-logical Centre of Russia is to "investigate the atmosphere-ocean-land system for the purposes of hydrometeorological forecasting" as well as acting as a "provision of the population policy makers and natural economy with operatio-nal hydrometeorological information, including warnings on adverse and disastrous weather phenomena" ("Hydrometeorological Research Centre of Russian Federation (Hydrometcentre of Russia)," n.d.). The Russians place great sig-nificance on research, focusing on areas of weather forming processes, modeling and mon-itoring atmospheric circulations, oceanic pro-cesses, and investigating the atmospheric and oceanic interaction. The Hydrometeorological Centre of Russia consists of 18 departments and laboratories with 11 administrative and management branches ("Hydrometeorological Research Centre of Russian Federation (Hydro-metcentre of Russia)," n.d.). Just like the Chinese, the Hydrometeorological Centre of Russia has played a large role in the satellite development during this time as well (see Section 9.5.3).

9.3.7 Mexico National Weather Service

Mexico's National Weather Service focuses on monitoring the atmosphere to identify weather events that can make an impact on the econ-omy and human life, all while collecting national climate information. Its Synoptic Network is composed of 79 meteorological observatories that observe and transmit real-time weather information, data, and conditions. This Synoptic Network includes 16 radiosonde stations along with 13 weather radars which became operational in 1993, covering most of the Mexican territory ("National Weather Service History: Translation," n.d.).

9.4 Radar Development

The advancement of weather radar after 1980 occurred worldwide (Figure 9.3). By 1989, there were over 100 different countries that housed weather radar, with over 600 total weather radars in operation. The main focus of these radars was to monitor and warn for flash flood events, severe weather events, wind events, and whether hail or tornadoes were present in thun-derstorms. Thirty-five percent of the world's radars were installed between 1980 and 1989 (Collier and Chapuis, 1990, p. 6). The moderni-zations of the major weather departments increased these totals and increased the quality of the radars during the 1990s through the 2000s. Most of today's 877 radars connected with the WMO can be found across the United States, Europe, Brazil, Mexico, Canada, Russia, and South Africa. Of the countries with 10 or more radars within the WMO, the United States leads the way with 220, followed by Australia with 66, Brazil with 44, the Russian Federation with 36, and Canada with 31 ("Number of Countries Radar," 2015). For a full table of the WMO radar system, please see Appendix Figure B.

9.4.1 Radar in Europe

In 1984, a five-year project called Cost 73 was put into action to further the establishment of operational weather radars across several European nations. The goal of this project was to make a more harmonious system by sharing information and data, and to maximize the benefits of radar technology. An issue that was not corrected with this project, however, was the type of radar. Each radar installed was not the same in all the other countries, leaving no standard for Europe (Newsome, 1992, pp. 2–3). Projects such as Cost 73 as well as others prior to and after were aimed to develop radar systems and precipitation measuring techno-logies, and to improve telecommunication, nowcasting, and weather modifications. But the most crucial aspect of these radar projects

WMO radar locations worldwide

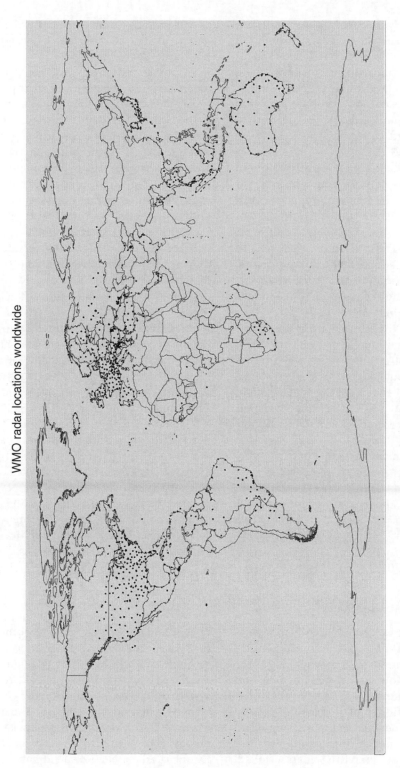

Figure 9.3 The World Meteorological Organization (WMO) is a worldwide organization with 185 member countries and six territories. The map displays the locations of a l WMO radar systems worldwide, with the vast majority located in the United States, Europe, coastal Australia, and the southeast coast of Asia. *Source*: World Meteorological Organization.

was the overall advancement in a shared network of advanced weather radars throughout Europe (Collier, 1992, p. 3). Weather radar is the only way to measure the location and intensity of precipitation in real time, whether being rain, snow, sleet, or even hail. No other method or computer allows as accurate a measurement and interpretation as radar ("Rainfall Radar," 2012).

Weather radar data are fed directly into the NWP models. These data come from radar networks, such as the non-stop radar network in the United Kingdom and Ireland that finishes each scan in five minutes all at different elevation angles. These data are then sent to the Met Office headquarters to be processed, and a picture is then created. This radar network has a resolution of up to 1 km (0.6 miles). The picture is further cleared up by Met Office staff, who remove the "clutter" from any reflection off of hills, buildings, or even birds and insects. Some "clutter" also occurs when radars are under strong high-pressure systems. Another aspect of radar control that needs to be adjusted at times is during winter weather events and with the rain–snow line transitions. Radar returns may show up as strong echoes in this type of circumstance, which may be interpreted as heavy precipitation, when in fact these bright returns simply show the transition of snow to sleet and rain. This reflection is often referred to as "bright banding," and occurs not only in Europe but also anywhere storm systems produce mixed precipitation ("Rainfall Radar," 2012).

9.4.2 Radar in the United States

In the United States, the NEXRAD modifications took place around the turn of the century (as stated in Section 9.1.2). The main part of the NWS modernization project was this new and advanced standardized NEXRAD radar network. The top priority of this system was to better the severe weather warning systems and to increase lead-times during extreme weather events. There were 165 NEXRAD radars installed, as well as many other types of radar that were and still are operated by television stations ("Weather Radar," 2015).

The most recent advancements to weather radar came in the form of dual-pol Doppler radar. This new technology was a modification to the NEXRAD radars, creating a better representation of precipitation type, particle size, and intensity. This modification began in 2011 and ran through 2013. The main difference between a regular Doppler radar and dual-pol radar is that dual-pol radars collect data on both a horizontal and vertical level, collecting the often-missed characteristics and data in weather events. This radar creates a two-dimensional view rather than the simple one-dimensional view of a single Doppler radar (Figure 9.4). The most significant achievement of the dual-pol was in regards to detecting weather types, more accurately showing how much precipitation is actually hitting the ground in flooding events, identifying debris in the air during tornadoes, and detecting a more accurate depiction of hail size and location. This modification cost the NWS $50 million, but is predicted to save the nation $700 million per year by increasing preparation and safety, and by reducing damage related to weather events ("Dual-Polarization Radar Training for NWS Partners," 2015). After the NWS implemented the dual-pol radar into its networks, Canada and many other European nations followed, modifying their radars to gather the same benefits as the United States.

9.4.3 Radar in Asia and Australia

These radar advancements did not just occur in Europe and the United States. Another location that has seen a large upgrade in weather radar technology from the 1990s through 2013 is in Southeast Asia. Indonesia has updated its radar system by changing from the more conventional original radar to the Doppler radar in 2006. Malaysia has developed its weather radar technology into a network that consists of 12 Doppler radars. New Zealand has a network of

Conventional radar

Dual-polarization radar

Figure 9.4 The conventional radar was revolutionary to the field of meteorology when it was first put into use. Since then, advancements in technology and science as a whole have led to the development of the dual-pol radar. The figure demonstrates the difference between the two styles of radar and how the dual-pol radar gets a better and more accurate depiction of the area by sending pulses out in both horizontal and vertical directions. This leads to a two-dimensional picture as opposed to the original one-dimension. *Source*: Courtesy of NOAA Photo Library.

weather radars covering nearly the entire region, with eight Doppler radars scanning. The recent upgrades to Australia's impressive radar network have been greatly beneficial, but do not fix the issue of coverage, as most of this radar network covers almost exclusively its coastal locations (Ibrahim, Van Dijk, and Stringer, 2013). A large hole is found in the center of the country where a much less densely populated area exists. Because few people are located in this area of the country, the need for radar is limited. This, however, detracts from the meteorologist's capabilities of observing what is occurring in that "hole," further limiting the ability to forecast and advise those who do happen to live in this area and the surrounding regions as well (Lane, Atkinson, and Stringer, 2017). New Zealand and Australia also have a radar data

sharing agreement, one of the very few data sharing agreements in this region of the world. This agreement was created to fix a similar issue to that of the "hole" in central Australia. The proximity of these two nations to each other, and the frequent weather systems that move between them, calls for the need to share data in order to be able to follow weather events more clearly and accurately. Most of the other nations in this region do not share their radar data. This is because most of the areas are remote enough that there is no need to see what is happening in other regions through the use of radar. Twenty-two countries and territories in total in the west Pacific are equipped with at least one radar, including the Philippines with six, Fiji with three, and Hawaii itself having four (Ibrahim, Van Dijk, and Stringer, 2013).

9.5   New Satellite Developments

Weather satellites are often thought of as the most important tool available to forecasters (Figure 9.5). A satellite is the only tool that sees a storm or weather system from above. Satellites can measure countless variables in the atmosphere, storms systems, ground, oceans, and more. While many nations have successfully launched and operated weather satellites since the 1960s (see Chapter 5, Section 5.5), there has been a huge expansion in satellite coverage from the 1980s through the early 2000s. These advanced weather satellites greatly monitor the world every second of every day and provide weather forecasters and researchers with the information they need to keep the world safe and to grow the knowledge of the atmosphere and environment.

9.5.1   United States' GOES and TIROS Satellites

In 1980, the United States continued its satellite growth by launching the GOES-4 satellite into orbit. This satellite was of major significance, as

it became the first geostationary satellite to provide continuous vertical profiling of the temperature and moisture of the atmosphere. By 1983, the international community saw a valuable advancement when the International TOVS, TIROS Operational Vertical Sounder, was put into use. This optimized and standardized the processing procedures so accurate data could be shared and made available universally. More advancements came in 1994 when the GOES-8 was launched, followed by the GOES-11 in 2000, and the GOES-14 in 2009. In 2010, the NOAA (National Oceanic and Atmospheric Administration) introduced a new technology that created the weather satellite data to be displayed in two and three dimensions. The GOES satellite series is part of the NOAA, but is often shared not only with the NWS but also with the public, television, radio, Internet, weather services, other countries, the space agency, and other private sector business (Anderson, *et al.*, n.d.).

As stated, the GOES satellite series is considered geostationary. That is one of the two different types of weather satellites. GOES stands for Geostationary Operational Environmental

Figure 9.5 The difference in technology from 1970 to today is extremely dramatic. One example of how far we have come since 1970 is when looking at satellite images and quality. To the left is an image from the 1970s from the TIROS 1 satellite of the Gulf of St. Lawrence and the St. Lawrence River. The image on the right was taken in 2013 from the Suomi NPP satellite of New England. *Source*: Courtesy of NOAA Photo Library. (*See color plate section for the color representation of this figure.*)

Satellites. Geostationary satellites are launched into orbit, circling earth at the equator at the same rate of speed the earth rotates. This orbit is a specific distance from earth and is called a "geosynchronous orbit." These types of satellites are great for short-range forecasting and now-casting. They continuously monitor the same location, staying over the same spot at all times. These satellites help pick up information crucial to detect and forecast tornadoes, flooding, hurricanes, precipitation estimates, snow pack and cover, wild fire, and more. As discussed in Section 5.5, the first GOES satellite was launched into orbit in 1975. Today, GOES-13 and GOES-15 are operating for the United Sates. GOES-14 is in orbit as standby in case of any failure or malfunction of the other satellites used by the United States ("NOAA's Geostationary and Polar-Orbiting Weather Satellites," 2014).

The second type of satellite is the polar orbiting satellite. Polar orbiting satellites are very useful for long-range forecasting. This is because they are not fixed over the same location, and instead are launched into orbit on a constant circular path around the earth's poles. As the satellite orbits, the earth rotates. This creates a path for the satellite to scan the entire globe, all the while keeping the same exact path itself. The two polar satellites used by the United States are known as "advanced television infrared observing satellites," TIROS-N or ATN. These two satellites create a uniform data field for not only the United States but also the entire world. This is where their role for longer-range forecasting comes into play. The pictures created by polar satellites show the overall advancement and state of the atmosphere across all locations. Polar orbiting satellites also cover the poles, which are not observed by the geosynchronous satellites due to the curve of the planet. Polar orbiting satellites are not useful, however, for nowcasting and active weather events. This is because the satellite does not stay overhead of a specific location long enough to show progression. The satellites are set to cross the equator twice a day, once at 7.30 a.m. and once at 1.40 p.m. local times. This allows for data to be updated every six hours on any location on earth ("NOAA's Geostationary and Polar-Orbiting Weather Satellites," 2014).

9.5.2 Europe's METEOSAT

The METEOSAT and METEOSAT Second Generation are the series of geosynchronous satellites placed into orbit over Europe. These satellites are operated by EUMETSAT, a global operational satellite agency in Europe. The layout of the satellites from this series creates coverage over Europe, Africa, and the Indian Ocean. Currently, METEOSAT-8, 9, and 10 are all of the next-generation satellites, replacing the original satellites of the 1980s and 1990s. These satellites operate over Europe and Africa, with METEOSAT-10 as the prime satellite of the series. METEOSAT-7 is of the older generation satellite series, but is operational over the Indian Ocean. Europe is also in charge of two polar orbiting satellites. In 2006, the METOP-A polar orbiter was launched, followed by the METOP-B in 2012 ("Operating Satellites," n.d.).

9.5.3 Russia's METEOR

The polar orbiting series of satellites for the Federation of Russia is called the METEOR. The series began in 1969 with the METEOR 1-1, which just recently fell out of orbit and into Antarctica in 2012. The METEOR 2 series began in the mid-1970s, with the METEOR 2-21 as the last of the second-generation METEOR satellites to be launched into orbit. This satellite is only useful for the winter part of the year due to decreased signal strength in the summer months. For this reason, the satellite is considered the winter satellite. The METEOR 3 series then began in the 1980s, with the METEOR 3-5 launched into orbit in 1991 and is only operational in the summer months, making it the summer satellite. The METEOR 3M is an advanced polar orbiting satellite series, with the METEOR 3M-N1 being launched in 2001, but never transmitting data successfully. In addition to these satellites, the Russians also have OKEAN4 and SICH1, which are radar imaging satellites with reception only within Europe ("Summary of Russian Meteorological METEOR Satellites," 2015).

9.5.4 United States' LANDSAT

The LANDSAT satellite series is a high-resolution earth observation satellite managed by NASA (the National Aeronautics and Space Administration), and the USGS (the United States Geological Survey). In 1999, the LANDSAT 7 was launched and followed up with the LANDSAT 8, which consisted of improved sensors and technology. In 2023, the LANDSAT 9 is set to be launched and operational. The LANDSAT series is important because it "represents the world's longest continuously acquired collection of space based moderate resolution land remote sensing data" ("Landsat Missions Timeline," 2015). In the four decades the LANDSAT series has been operational it has become a great resource for agricultural businesses and research, geology, forestry, regional planning and policymakers, education, mapping, and overall global climate change research ("Landsat Project Description," 2013).

9.5.5 Canada's RADARSAT

Canada's radar series began in 1995 with the launching of RADARSAT-1, Canada's first satellite launched into orbit. This began a program that had international impacts. The RADARSAT was used internationally to manage and to monitor earth's resources and to follow climate change across the world. The success that Canada had with its first satellite led to the 2007 launching of RADARSAT-2. This satellite was the next-generation model, updating on what was learned from the first satellite, and continuing the success in global scale information and data gathering ("Archived: Canadian Satellite RADARSAT-1 Celebrating," 2010).

9.5.6 Japan's GMS and MTSAT

Japan's Geostationary Meteorological Satellite (GMS) series consists of geostationary orbit satellites which are used by the WMO as part of the World Weather Watch. The GMS series relied heavily on the model and design of the GOES series. In 1977, the GMS-1 was launched and operational. This was followed by the GMS-2 in 1981. The GMS-3 was launched in 1984, followed by the GMS-4 in 1989, and finally the GMS-5 in 1995. After the GMS-5, Japan unsuccessfully launched the MTSAT in 1999. With a lack of satellite coverage following the failed attempt of the MTSAT and the decommissioning of the GMS-5, the GOES-9 was put into orbit in 2003 ("Geostationary Meteorological Satellite (GMS)," 2015). Japan then successfully launched its MTSAT-1R in 2005, followed by the MTSAT-2 in 2006, which added new technology to be able to detect fog at night as well as consisting of one visible satellite channel and three infrared satellite channels ("MTSAT (Multifunctional Transport Satellite) Himawari," 2015) (Figure 9.6).

Chapter Summary

While certain nations led the way during the 1980s through 2013 in advancements in NWP, weather models, weather departments, radar, satellites and other technologies, the world as a whole saw that advancements across the board were needed in order to keep up with technology, meteorological knowledge, and the need to protect every human on earth from weather-related disaster. The United States became a leader in radar and satellite technology, the Met Office became a world provider of one of the most recognized meteorology training schools, and countries such as Brazil, Australia, and South Africa began to see their place among the top weather research and forecasting departments in the world. There are many other nations than those described in this chapter, all of which saw a great modernization of weather agencies. Along with this, there were many other nations which experienced an increase in radar technology, such as Brazil and the Middle East, and increase in satellite technology,

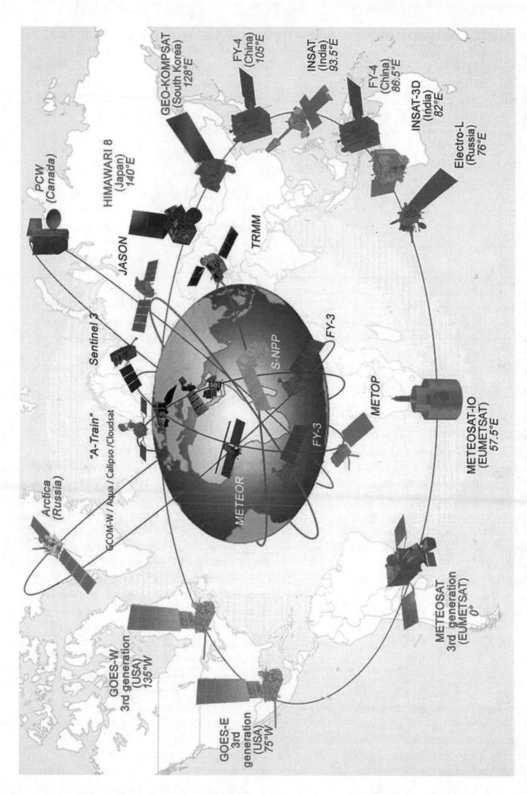

Figure 9.6 This image demonstrates the satellite traffic orbiting earth, with meteorological and climatological information and data being collected around the clock. Whether satellites are from the United States, Canada, Russia, Japan, South Korea, China, India, Europe, or other cooperative agencies, there are constant eyes on earth furthering the field of meteorology. *Source:* Reproduced with permission of WMO.

such as India, Pakistan, China, and South Korea. The world was in a meteorological boom during the 1980s through the turn of the century. The sharing of all the global data among these top governmental agencies and nations can be seen in world organizations such as the WMO and the NMHS, both often seen trying to help the lower income nations and territories, continuing the drive to protect as many people and livelihoods as possible.

References

Book References

Collier, C.G. (1992) *International Weather Radar Networking: Final Seminar of the COST Project 73*, Commission of the European Communities, Luxembourg.

Collier, C.G. and Chapuis, M. (1990) *Weather Radar Networking: Seminar on COST Project 73*, Commission of the European Communities, Luxembourg.

Newsome, D.H. (1992) *Weather Radar Networking: COST 73 Project/Final Report of the Management Committee*, Commission of the European Communities, Luxembourg.

Rogers, D.P. and Tsirkunov, V.V. (2013) *Weather and Climate Resilience: Effective Preparedness through National Meteorological and Hydrological Services*, International Bank for Reconstruction and Development/World Bank, Washington DC.

World Bank. (2008) *Weather and Climate Services in Europe and Central Asia: A Regional Review*, International Bank for Reconstruction and Development/World Bank, Washington DC.

Journal/Report References

Ibrahim, K., Van Dijk, W., and Stringer, R. (2013) *Current Status of Weather Radar Data Exchange*, World Meteorological Organization, Commission for Basic Systems, Workshop on Radar Data Exchange, Exeter.

Lane, A., Atkinson, R., and Stringer, R. (2017) *The Australian Weather Radar Network: Current and Future Challenges*, Surface Based Observations Section, Australian Bureau of Meteorology, Melbourne, Victoria, https://ams. confex.com/ams/pdfpapers/123509.pdf. 14 January 2017.

Website References

"About Us." (2015) *About Us: South African Weather Service*, SAWS, http://www.weathersa. co.za/about-us/about-us, accessed 15 September 2015.

Anderson, B., Hedges, L., Merchant, K., and Phillips, J. (n.d.) *Satellite Meteorology History: A View from Madison*, Space Science and Engineering Center, University of Wisconsin-Madison, http://www.ssec.wisc.edu/news/media/2012/07/foldoutV2.pdf, accessed 15 September 2015.

"Archived: Canadian Satellite RADARSAT-1 Celebrating." (2010) *News Release*, Government of Canada, http://news.gc.ca/web/article-en.do?nid=570879, accessed 15 September 2015.

"AWIPS Advanced Weather Interactive Processing System." (n.d.) *National Weather Service New York, NY Tour AWIPS Page*, NWS & NOAA, http://www.weather.gov/okx/Tour_AWIPS, accessed 15 September 2015.

"Brief History." (2009) *KMA Introduction*, KMA, http://web.kma.go.kr/eng/aboutkma/briefhistory.jsp, accessed 15 September 2015.

"Dual-Polarization Radar Training for NWS Partners." (2015) *Warning Decision Training Division*, NWS & NOAA, http://www.wdtb.noaa.gov/courses/dualpol/Outreach/index.html, accessed 15 September 2015.

"Geostationary Meteorological Satellite (GMS)." (2015) *Space*, Global Security, http://www.globalsecurity.org/space/world/japan/gms.htm, accessed 15 September 2015.

"History." (n.d.) *About Us: Japan Meteorological Agency*, JMA, http://www.jma.go.jp/jma/en/Background/history.html, accessed 15 September 2015.

"History of the Met Office College." (Updated 2015) *Met Office*, Met Office, http://www.metoffice.gov.uk/training/college/history, accessed 15 September 2015.

"History of the National Weather Service." (n.d.) *National Weather Service*, NWS & NOAA, http://www.weather.gov/timeline, accessed 2 July 2015.

"Hydrometeorological Research Centre of Russian Federation (Hydrometcentre of Russia)." (n.d.) *About Us: Hydrometeorological Centre of Russia*, Federal Service for Hydrometeorology and Environmental Monitoring, http://wmc.meteoinfo.ru/about, accessed 15 September 2015.

"IFPS/GFE Interactive Forecast Preparation System Graphical Forecast Editor." (n.d.) *National Weather Service New York, NY Tour AWIPS Page*, NWS & NOAA, http://www.weather.gov/okx/Tour_GFE, accessed 15 September 2015.

"International Role." (Updated 2013) *Met Office*, Met Office, http://www.metoffice.gov.uk/about-us/what/international, accessed 15 September 2015.

"Landsat Missions Timeline." (2015) *Landsat Missions*, USGS, http://landsat.usgs.gov/about_mission_history.php, accessed 15 September 2015.

"Landsat Project Description." (2013) *Landsat Missions*, USGS, http://landsat.usgs.gov/about_project_descriptions.php, accessed 15 September 2015.

"Message of Administrator." (2011) *About CMS*, China Meteorological Administration, http://www.cma.gov.cn/en2014/aboutcma/, accessed 15 September 2015.

"Mission & Vision." (2009) *KMA Introduction*, KMA, http://web.kma.go.kr/eng/aboutkma/mission.jsp. 15 September 2015.

"MTSAT (Multifunctional Transport Satellite) Himawari." (2015) *Space, Global Security*, http://www.globalsecurity.org/space/world/japan/mtsat.htm, accessed 15 September 2015.

"National Weather Service History: Translation." (n.d.) *Comision Nacional Del Agua, Servicio Meteorologico Nacional*, CONAGUA, https://translate.google.com/translate?hl=en&sl=es&u=http://200.4.8.20/es/smn/historia&prev=search, accessed 15 September 2015.

"NOAA's Geostationary and Polar-Orbiting Weather Satellites." (2014) *National Environmental Satellite, Data, and Information Service*, NOAA Satellite Information System, http://noaasis.noaa.gov/NOAASIS/ml/genlsatl.html, accessed 15 September 2015.

"Number of Countries Radar." (2015) *WMO Radar Database*, Operated by Turkish Meteorological Service (TMS), http://wrd.mgm.gov.tr/statistics/countries.aspx?l=en, accessed 15 September 2015.

"Operating Satellites." (n.d.) *What We Do: Operating Satellites*, EUMETSAT, http://www.eumetsat.int/website/home/AboutUs/WhatWeDo/OperatingSatellites/index.html, accessed 15 September 2015.

"Organisational Structure." (2015) *About Us: Bureau of Meteorology*, Australian Government, http://www.bom.gov.au/inside/org_structure.shtml, accessed 15 September 2015.

"Overview." (2015) *Overview: South African Weather Service*, SAWS http://www.weathersa.co.za/about-us/overview, accessed 15 September 2015.

"Overview of the Met Office." (Updated 2012) *Met Office*, Met Office, http://www.metoffice.gov.uk/news/in-depth/overview, accessed 15 September 2015.

"Public Weather Service." (Updated 2015) *Met Office*, Met Office, http://www.metoffice.gov.uk/about-us/what/pws, accessed 15 September 2015.

"Rainfall Radar." (2012) *Met Office*, Met Office, http://www.metoffice.gov.uk/learning/science/first-steps/observations/rainfall-radar, accessed 15 September 2015.

"Summary of Russian Meteorological METEOR Satellites." (2015) *National Environmental Satellite Data, and Information Service (NESDIS)*, NOAA Satellite and Information Service, http://noaasis.noaa.gov/NOAASIS/ml/meteor.html, accessed 15 September 2015.

"The Met Office in Southeast Asia." (Updated 2014) *Met Office*, Met Office, http://www.metoffice.gov.uk/about-us/what/international/southeast-asia, accessed 15 September 2015.

"Weather Radar." (2015) *NWA Remote Sensing Committee*, National Weather Association, http://www.nwas.org/committees/rs/radar.html, accessed 15 September 2015.

Figure References – In Order of Appearance

"NEXRAD and TDWR Radar Locations." (2016) *Radar Operations Center – NEXRAD WSR-88D*. NOAA. NWS, https://www.roc.noaa.gov/wsr88d/maps.aspx, accessed 20 July 2016.

"AWIPS Advanced Weather Interactive Processing System." (n.d.) *National Weather Service New York, NY Tour AWIPS Page*. NWS & NOAA, http://www.weather.gov/okx/Tour_AWIPS, accessed 15 September 2015.

"IFPS/GFE Interactive Forecast Preparation System Graphical Forecast Editor." (n.d.) *National Weather Service New York, NY Tour AWIPS Page*. NWS & NOAA, http://www.weather.gov/okx/Tour_GFE, accessed 15 September 2015.

"WMO Radar Database." (2016) *WMO Radar Database*, http://wrd.mgm.gov.tr/default.aspx?l=en, accessed 22 July 2016.

"Dual Polarization." (2012) *Radar Operations Center, NEXRAD WSR-88D*. NWS & NOAA, http://www.roc.noaa.gov/WSR88D/DualPol/Default.aspx, accessed 15 September 2015.

"Happening Now: NOAA operations in real time." (2014) *NOAA Satellite and Information Service*, http://www.nesdis.noaa.gov/fourbox/03-25-13/, accessed 22 July 2016.

"Global Planning." (n.d.) *Global Planning-Space-based GOS*. World Meteorological Organization, www.wmo.int/pages/prog/sat/globalplanning_en.php, accessed 25 July 2016.

10

Meteorological Agencies at the Global and Private Level

Kevin Anthony Teague

The establishment of meteorological agencies is essential for helping us understand atmospheric elements and processes. There are numerous meteorological agencies throughout the globe that are individualized to provide services to one specific nation; however, there are also global meteorological agencies that provide services on an international scale, similar to the role of the United Nations. Each agency is a part of a greater purpose: to advance the understanding of the atmospheric and meteorological processes.

10.1   World Government Agencies

In order to have a consistent global data set and monitoring system, an establishment of a multinational organization is needed. World agencies created for meteorological purposes originated in the late nineteenth century. The first organization formed for the purpose of exchanging meteorological information among countries of the world was the International Meteorological Organization (IMO) (1873–1951). Meetings of the IMO consisted of paying attention to the standardization and optimization of weather observing and reporting practices and the needs of research workers. Discussions also took place regarding the development of committees, such as a maritime and aeronautics committee.

This organization had many worldwide scenarios to overcome which enabled changes in the committee such as deactivation and reactivation when necessary. During World War I, the IMO ceased operations of the committee and international meteorological developments lay dormant. This did not mean that meteorological discussions and advancements stagnated. Allied nations developed their own structures of the IMO and meetings continued. Aviation rivalries occurred immediately during the war, demanding data and information about visibility, cloud heights, and wind. After World War II, new principles were adopted and a new committee was assembled for the IMO.

The United Nations, an intergovernmental organization that promotes international cooperation, created an organization that specifically handles environmental issues in 1947, and the World Meteorological Organization (WMO) has been operating ever since. The WMO is a specialized agency of the United Nations that speaks on the state and behavior of earth's atmosphere and its interactions with the oceans. The organization also examines and studies the climate, the atmosphere, and the distributions of water resources. The development of meteorology and hydrology with their practical operational applications is undertaken by the WMO. Many different roles are played by the organization in its efforts to reduce loss of life and to monitor and protect the environment.

The Evolution of Meteorology: A Look into the Past, Present, and Future of Weather Forecasting, First Edition.
Kevin Anthony Teague and Nicole Gallicchio.
© 2017 John Wiley & Sons Ltd. Published 2017 by John Wiley & Sons Ltd.

The WMO collaborates with several different programs as well to enhance knowledge on specific sections of meteorology and hydrology. International efforts are extremely important in collaborating research and necessary actions on a global scale. One of the principal roles of the organization is to provide the public, and more specialized users, with reliable and accurate weather products and services. "These products and services help to ensure the safety and protection of people and of goods, at times of occurrence of extreme weather phenomena, but also assist in the efficient and economic pursuit of business and of the domestic activity of citizens engaged in their own particular area of pursuit and responsibility" (Fleming, 2005, p. 3). WMO innovative weather products and services need to evolve with the ever-new and variable weather parameters as time goes on. They must be communicated and distributed in a timely manner to users and the public in order for them to be efficient and used for their purpose. The National Meteorological and Hydrological Services (NMHS) has broadcasting guidelines in order to fully support users and its services are readily available to users and the public (see Chapter 9, Section 9.3). There are many factors that go into preparing the weather for broadcast, and the WMO thoroughly ensures that it delivers exact, accurate, organized, and understandable information.

The United Nations has several organizations that relate to meteorology. The United Nations Office for Disaster Risk Reduction (UNISDR) specializes in climate change adaptation, building disaster resiliency, socio-economic and humanitarian fields, and strengthening international systems. Work is conducted through this organization all around the world, with assemblies in the United States, Africa, Asia, the Arab states, and Europe (Sarukhanian and Walker, n.d.). The assembly works with many other organizations, institutions, parliaments, platforms, companies, and the media in coordination efforts to better their causes. Disaster risk reduction (DRR) aims to reduce the damage caused by natural hazards caused by meteoro-logical processes. Disasters often follow natural hazards and the severity depends on how much impact the hazard has on society and the environment. DRR is the concept and practice of reducing disaster risks through systematic efforts to analyze and reduce the causal factors of disaster ("What We Do," n.d.). DRR efforts are taken from a global standpoint in preparation for humanitarian needs. Earth's natural climate poses risks to human life at times and it is necessary to have global precautions to protect human and animal existence.

The National Center for Atmospheric Research (NCAR) is a federally funded research and development center devoted to service, research, and education in the atmospheric and related sciences. The mission of the NCAR is to understand the behavior of the atmosphere and related physical, biological, and social systems ("About NCAR," 2015). The NCAR is federally funded by the United States; however, it endeavors to communicate both nationally and internationally and to operate on a global scale. Its research covers a vast array of topics, primarily in the areas of investigation of atmospheric chemistry, climate, weather science and hazards, interactions between the sun and earth, computer science innovation, and effects of weather and climate on society and national security. The NCAR takes action to help the environment, along with educating the public on atmospheric-related sciences. Scientific visionaries designed a place where researchers could go to interact and share expertise, ideas, and knowledge with their university colleagues. The University Corporation for Atmospheric Research (UCAR) was established to support the scientific mission that the NCAR upholds. A university devoted to atmospheric research allows for a concentrated education and knowledge base for students and employees.

Based in the United States and operating on a global scale, the National Weather Association (NWA) and the American Meteorological Society (AMS) aid in supporting excellence in operational meteorology. Members in these societies are meteorologists and may be granted a seal of approval from the society. Collaboration of scientists ensures that society benefits from the most current scientific knowledge and understanding available. Both societies provide global leadership in the fields of atmospheric science. Journals are published and distributed by these organizations, keeping their members educated and updated. "The American Meteorological Society (AMS), founded in 1919, is the nation's premier scientific and professional organization promoting and disseminating information about the atmospheric, oceanic, and hydrologic sciences" ("About AMS," 2015). "The National Weather Association is a member-led, all inclusive, non-profit, professional association supporting and promoting excellence in operational meteorology and related activities since 1975. Members have many opportunities to share information, news, studies, and concerns related to operational meteorology and related activities through committee work, submitting correspondence or articles to NWA publications such as the Newsletter and the Journal of Operational Meteorology, making presentations or leading workshops at the Annual Meetings, helping to maintain and add information to the NWA website, and to network with great people in a wide variety of careers. Members work together on strengthening education and training for the general public, students and users of weather data" ("About the National Weather Association," 2015). Meetings are held throughout the year, along with an annual conference. These meetings and conferences enable networking to take place, along with distribution of valuable knowledge.

"The Royal Meteorological Society, RMS, is the professional and learned society for weather and climate. The Society serves not only those in academia and professional meteorologists, but also those whose work is affected in some way or another by the weather or climate, or simply have a general interest in the weather. The membership includes scientists, practitioners, and a broad range of weather enthusiasts" ("About Us," n.d.). The programs of the RMS are broad and diverse. The RMS publishes four international science journals and an in-house journal. Its program of conferences and meetings brings together scientists from across the international community, along with having a number of very active local centers and special interest groups. The RMS was established on 3 April 1850 as the British Meteorological Society and was incorporated by Royal Charter in 1866, when its name was changed to the Meteorological Society. The privilege of adding "Royal" to the title was granted by Queen Victoria in 1883. The merging with the Scottish Meteorological Society took place in 1921. The patron of the Society is the Prince of Wales.

The Meteorological Society of Japan and the Japan Meteorological Agency (JMA) were both founded in the late 1880s. The Meteorological Society of Japan is active in meteorological research and promotes the progress and development of such research in cooperation with related academic societies both in Japan and around the world ("An Introduction to the Meteorological Society of Japan," 2008). Meteorological services in Japan were initiated in 1875 by the Tokyo Meteorological Observatory, now known as the JMA. Now, the JMA serves as one of the most advanced and leading national meteorological services in the world. The agency assumes both national and international responsibilities. There is an implemented Meteorological Service Act in place that maps out ultimate goals, such as prevention and mitigation of natural disasters, safety of transportation, development and prosperity of industry, and improvements of public welfare ("History," n.d.). Both organizations are engaged in international partnerships in order to monitor earth's environment. It holds annual and regularly scheduled meetings, and sends out publications and conducts research in meteorological advancements on a global scale. These different actions allow for these associations to be at the leading edge of meteorological knowledge, technologies, and advancements.

10.2 Operations of World Organizations

In order for world and national organizations to operate efficiently and successfully certain procedures are necessary. The WMO is not the only organization that is concerned with meteorological processes and their climatological interactions on a global scale. Many nations have their own programs that conduct research and develop global meteorological issues. Just as any other organization, there are pros and cons to smaller global programs. Large-scale organizations have the advantage of a greater access to data and technologies. Smaller programs have the advantage of a concentrated work environment, allowing for detail-oriented projects. Organizations, large or small, pros or cons, will aid in advancements and better understanding of meteorological and atmospheric processes. However, credibility is a key component of these national meteorological services. Many organizations are keen to improve their status. Credibility lies on the accord of the nation in which the services are occupied. And assessment by the public, governments, and other agencies, along with the results of high levels of performance, lends credibility to these organizations. Success comes from several different factors. One factor is the success to reach a wide base of the public. Communication is a necessity, and it must be in a language and terminology that listeners can understand, in order to properly take action in certain situations. The delivery of weather information is important and must be presented to the audience in a way that influences the user, along with topics that develop the focus of the viewers. Timely delivery is important and must be done through mass media. It is imperative for these organizations to have strong connections with the media. They must collaborate and have a partnership in order to present information to the public in a respectable and proper timeframe for action to take place. The quality of communication with the public does not matter if the service does not provide accurate and viable data and information. Strong business training and performance are necessary for these organizations to be upheld. It is imperative for advancing in these services to have employees able to grow, learn, and educate. Society's expectations of the meteorological community have increased over recent decades and organizations are expected to deliver a suite of traditional and new products and services with timeliness, accuracy, and quality higher than ever before ("Credibility of National Meteorological Services," n.d.). The global pressures to create advancements in meteorological forecasting and technologies have only continued to grow. Large organizations are especially critiqued for their movements and furthering of knowledge of the atmosphere and meteorological elements and processes.

10.3 Global Discussions

International cooperation provides a multiplicity of meteorological information available for use on a global scale. Global interactions and sharing require an outsourcing of advancements, filtered through numerous aspects such as conferences, discussions, seminars, meetings, media, and the Internet. Topics of these notions include weather patterns, atmospheric stability and homeostasis, ecology, human involvement, and disaster matters. These holdings of interactions between societies promote expansion in the field of meteorology and atmospheric science. Scientists around the world have opportunities to showcase their new technologies and research. The projected outcome of these meetings is to create a larger understanding for weather and earth's atmospheric processes. In order for worldwide meteorological advancements, society has to work on a global scale to combine information, data, research, and new technologies. Organizations are set up to work internationally, with the intent to enhance numerous aspects of meteorology.

10.4 What a Private-Sector Weather Business Entails

When it comes to weather forecasting, global and governmental agencies are not the only suppliers of information. As the world became more commercialized, businesses needed every edge possible to better themselves when compared to their competitors. Expenses had to be minimized and there was an emergence of greater value placed on employee and public safety. One area that seemed to be of influence was and still is the weather. With a better understanding of weather conditions and forecasts for a specific business or a specific location, businesses were able to further advance themselves. Weather forecasts are always available to the public through the government, but many times there are more specific and more intimate details that need to be covered calling for the need to have a more personal and private approach. Meteorologists in the private industry are the ones responsible for this new weather-relaying business.

A private weather business is responsible for many different needs that their clients may request. A private weather business consists of professional meteorologists, meaning they must have a science degree or higher in atmospheric or related fields, or 20 semester hours of credit in atmospheric or related fields along with three years of professional experience, as per the AMS's guidelines (Spiegler, 2007, p. 11). The most significant changes within the private sector of meteorology occurred after the mid-1990s (Spiegler, 2007, p. 1). David Spiegler of the AMS relates the growth and importance of the private sector to the rise in weather-related catastrophes: "the recent number of extreme weather events, record-setting precipitation accompanied by extensive flooding, record numbers of hurricanes and damages, etc., will lead to continued steady, and perhaps accelerated, growth in the forensic meteorology segment" (Spiegler, 2007, p. 20).

Private weather businesses work on anything from small event forecasting, such as a wedding or school field trip, to bigger events such as professional sporting events or concerts. Private forecasters are responsible for site-specific weather forecasts for local farmers, for public education transportation, and weather-related delays or closures. They are also used for construction companies, energy companies, and even regular everyday businesses, such as Home Depot, Wal-Mart, and clothing industries. With these types of companies, trend forecasts are used to improve forecasts that may be needed by consumers down the road. Other areas in which a private meteorology business plays a valuable role is with stock market and energy trading. Understanding the trends of weather conditions can go a long way to predicting the bottom line of a company. Having a professional meteorologist on staff or purchasing the services of a private business can help brokers buy or sell stocks of companies at the perfect time to increase value and income. The same approach is used with energy trading. If a forecaster knows of a warm spell or a dry spell that will hit a region, traders will use that knowledge and buy or sell with the weather in mind. Commercials, sporting events, movies, TV shows, and much more also use the services of private meteorologists. With a "time is money" attitude of major importance to producers and owners, weather forecasts can help schedule video shoots or game start times in order to get the most out of the time allotted to them.

Another major field is marine forecasting. There are many private businesses that are in the specialized field of forecasting for large vessels and fleets that voyage across the oceans, transporting the world's cargo. Forecasters are required to supply specific forecasts to aid in the safety of the crew onboard the vessel, to provide the fastest mapped-out track to take, and to better the use of fuel consumption during the voyage. Marine forecasts go across international borders leaving the meteorologists with the challenging responsibility to understand the atmosphere and weather characteristics across the whole world, rather than in just a specific country or even a specific region or city within a country.

Private companies are also responsible for insurance claims in regards to weather-related damage or accident claims. This is often referred to as "forensic meteorology." Here a meteorologist has to research what happened, not forecast what will happen. It is a different skill set that a meteorologist in the forensic field has to develop, calling for extensive research into weather records and understanding the atmospheric conditions, radar and satellite images of the time in question, and much more. High-quality computer software and programs are often required to gather the information needed to enable clients to fight or defend their cases. Forensic work is valuable when it comes to aiding police or detective work in certain criminal cases. Weather can be a help or a hindrance during a criminal investigation. Certain clues may be found because of weather conditions, while other clues may be removed or deemed inadmissible in a court due to the effect of precipitation, temperature, and other weather factors. Oftentimes, police, fire, and city officials need the help of a trained meteorologist to provide the necessary information of what the weather was like at a specific time and location, as well as forecasting what the weather will be like during evidence collection and investigation.

10.5 Government vs. Private Sector

The private sector differs from governmental weather departments, although some countries and regions include the private sector in with the government as its own separate branch. Some of the larger countries in which there is the greatest complete separation of government and private-sector business are Australia, Canada, Japan, and the United States. These nations, especially the United States, make the difference known between governmental and private. They make most, if not all, of their data

and weather information freely accessible to the public. This access to the same information that the governmental departments use to make and distribute their forecasts allows for more private-sector growth, as the government is not the main player (House of Commons Select Committee on Science and Technology, 2012, p. 31). In the United States, taxpayers fund the public display of the NWS information, and the private sector then commercializes it. Nations such as those under the watch of the Met Office and many other offices in Europe and across the world do not have as free access to information, and therefore have far fewer separate private-sector meteorology businesses and more consumer type services straight from the Met Office or their respective governmental agencies. The free flow of information makes the United States the primary resource of free yet dependable information for the NMHS in the poorer regions of the world (World Bank, 2008, p. 31). The increased cost combined with the decline of economies and different philosophies about the role of federally funded weather services have continued the trend for many European nations to be partly privatized with commercial branches within main government departments (Committee on Partnerships in Weather and Climate Services, *et al.*, 2003, p. 20). In contrast, the NWS goes as far to say, "the NWS firmly believes that the private weather industry plays an important and essential role as a partner in ensuring that the Nation receives the full benefit of weather and hydrometeorological information for promoting protection of life and property, and economic prosperity" and "the NWS will not compete with the private sector when a service is currently provided or can be provided by commercial enterprises, unless otherwise directed by applicable law" ("Policy Statement on the Weather Service/Private Sector Roles," 1991). The United States and the NWS also allow for the spread of all their information and data internationally, while other countries and offices limit what is spread across borders.

While the private sector has taken off since the 1990s, it can be said to have first started to make appearances in the 1940s with TV and radio forecasts specifically detailed for farming communities. Today, due to ever-changing and sophisticated technologies and computers, the private sector sometimes outdoes even governmental departments. State-of-the art technology is easier to implement outside the government, often leading to better supercomputers and the newest weather programs and devices. A main complaint of governmental agencies is that they feel they should not lose out financially on the work and data that they collect and create, while the private sector takes that information and commercializes it. The other side feels that weather can be seen just as any other governmental service, such as police departments, where they are supplied by the government and taxpayers' money, but often work alongside or are coordinated with the private industries of alarm services and private securities. This example shows a private business using and working alongside a governmental service for profit (Smith, 1999). Many European countries as well as others do not allow total free access to their data, at times acting more as monopoly holders of meteorological data. They do not block all information but restrict some to their citizens and internationally (Aichholzer and Burkert, 2004, pp. 164–167).

Governmental agencies work extremely hard to provide a lot of information, detailed, accurate, and easily understandable, to large numbers of clients, all for different purposes and needs. Due to these large agencies, some information may be lost, hard to see, or just not understood by a layman or businessman. As previously mentioned, the private sector can take care of these issues personally. Private-sector organizations act as consultants for entities with specific needs. Instead of a large region or city, a private-sector meteorologist can take its time focusing on the microclimate of a specific location, measuring and predicting frost chances for a specific farm at a specific time on a specific field. This pinpoint and personalized request is more suited to the private sector than a government department. Private-sector organizations can better explain to the client what is forecast, adding the consulting touch that is not as accessible from the NWS. The private sector can in theory be looked at as exactly what the Met Office describes it as, as a branch of the main service. Whether that branch is still operated by the government, or whether it is fully private, that branch works with the government and creates its own interpretation and understanding.

10.5.1 The United States

The United States is the prime example of fully open resources and information paid for by the taxpayer. This relationship between the government and free enterprise comes from the US government and departments such as the National Oceanic and Atmospheric Administration (NOAA) believing that the nation as a whole benefits more when information from federal agencies and the private sector is shared. The NOAA "recognizes the government best serves the public interest by cooperating with private sector and academic and research entities to meet the varied needs of specific individuals, organizations, and economic entities" and will "take advantage of existing capabilities and services of commercial and academic sectors to avoid duplication and competition in areas not related to the NOAA mission" ("Policy on Partnerships in the Provision of Environmental Information," n.d.). There are currently hundreds of commercial weather providers and companies across the United States, with the NWS even listing the providers for public viewing and knowledge. Some of the fields of specialty that these providers represent include applied meteorological services, agricultural weather services, basic consulting, weather equipment/software manufacturer, forecasting and weather prediction, and marine weather. In 2008, the AMS exemplified the significance of the private sector by creating an extensive goals and objectives list focusing solely on the private sector. Some of its goals and objectives included topics such as making sure the AMS is kept abreast of current practices and trends in the private sector, to ensure it serves as a resource for the private sector, to encourage professional development, and to stimulate activities pertaining to the practicing of private-sector meteorologists. The AMS aims to raise awareness of the benefits of private-sector and private-sector mentor programs. It also encourages private-sector meteorologists to complete the Certified Consulting Meteorologist (CCM) program ("Goals and Objectives," 2013).

10.5.2 Australia

Australia is another major nation that follows an approach that is similar to the United States', but it is not as open-ended. Australia's Bureau of Meteorology has to provide data and products to many different users. This was agreed to in the 1974 Trade Practices Act, which "prohibits a person or an entity with significant market power from taking advantage of that power for anti-competitive purposes" (Gunasekera, 2004, p. 19). The 1997 agreement, called the Australian Competition and Consumer Commission (ACCC), helped produce a policy for general access to the services and products of the Bureau of Meteorology. Under this policy, users of these services are identified as, "national community, identifiable commercial users, civil aviation and defense forces and national meteorological services of other countries" (Gunasekera, 2004, p. 19). A difference from the US policies is that this policy also requires users to sign an access agreement, which states that users are generally precluded from using the data shared by the Bureau for the purpose of "issuing independent forecasts or warnings to the Australian public" (Gunasekera, 2004, p. 19). Internationally, the Bureau fulfills all the obligations of sharing data and information through the WMO, using WMO Resolution 40, which states that weather information is essential, and endorses its free and unrestricted exchange for the global public good (Gunasekera, 2004, p. 114).

10.5.3 Canada

Canada shows its cooperation with the private sector by also listing over 50 private-sector businesses under its own governmental website. These companies provide not only your everyday weather forecasting services but also include some worldwide technological and instrumental companies. These companies produce monitoring systems, aviation training and observing systems, specializations in the measuring of turbulence in natural waters, wind energy, and much more. Many of these companies focus on the technological and research side of meteorology ("Canadian Meteorological Private Sector," 2015).

10.5.4 Europe/United Kingdom

The commercialized meteorological industry in the United Kingdom, and much of Europe, is often assigned by the Met Office's commercial branch of services. Unlike the NWS, the Met Office specializes in consulting and risk assessment aspects of weather in the commercial sector as well as the government sector.

Although the Met Office does make a wide range of data and information available for re-use, which means, "the use by any person of information held by the Met Office for a purpose other than the initial purpose within the Met Office's public task for which the information was produced," there is a price for certain other information. The Met Office aims to keep the selling of this information fair, consistent, and non-discriminatory. The Met Office tries to ensure an equal playing field between the Met Office itself and the private sector ("Pricing Policy," 2014). The commercial side of the Met Office consists of four programs: consulting, transport, media, and training. The information available for free to the public and the private sector consists of basic weather forecasting and severe weather warning services. Some other information is made accessible without cost but must be done so through an agreement, while other information must be paid for ("Pricing Policy," 2014).

Not all of the private sector is under the control of the Met Office. One of the world's largest private-sector weather businesses is MeteoGroup, based in London but with branches stretching across the world. This company is an example of a large European-based commercialized weather business that has multiple sectors of business within it, ranging from energy to transport, marine, construction, insurance, retail, and others, just as with the numerous weather companies within the United States. While the businesses are not as widespread across Europe, they do exist and offer similar services as the US weather business, but with the Met Office having its own hold on the commercial industry, a stronger competition of services is created limiting the expansion of the private sector in that area ("Europe's Largest Private Weather Company Opens Office in Singapore," 2012).

Although companies like the MeteoGroup and other European private weather businesses exist, France has done things a little differently by separating the commercial industry from its Ministry of Transport, giving it its own federal entity and commercializing the generation of meteorological information. France then charges for phone and other hydrological and meteorological services, increasing the revenue of Météo-France (Cummings, Dinar, and Olson, 1996, pp. 14–15).

The Royal Meteorological Society of the United Kingdom is a professional weather and climate society, with members ranging from scientists, professionals and practicing meteorologists, to weather enthusiasts. This society, similar to the AMS, provides outlets for meteorologists to find private-sector opportunities for employment, and also for businesses influenced by weather to find private sector weather businesses for services ("Organisations," n.d.).

10.5.5 Japan/East Asia

In 1993, Japan took a different approach in the commercial industry by initiating a collaborative program with its own governmental department,

the Japan Meteorological Agency (JMA), and with 20 private meteorological companies. This agreement allowed the sharing of information to serve a market in need of weather information (Cummings, Dinar, and Olson, 1996, p. 14). This was done at marginal costs to users, and involved an abundance of weather-related data. Some of this information consisted of observational data from radar and Meteorological Terminal Aviation Routine Weather Reports (METAR), global forecast models, satellite imagery, short-range forecasts and nowcasts, analysis of forecast charts, advisories and warnings, marine services, earthquake and tsunami warnings, and atmospheric monitoring data. Some of the other type of data made available through this agreement included summaries of data, summaries of satellite imagery, and monthly and annual observation reports, analysis, and statistics (JMBSC, 2015, pp. 2–6).

10.6 The Lesser-Known Services of the Private Sector

As previously described, most topics discussed when talking about private meteorological companies involve weather-forecasting and data-providing services for other outlets. There are many other aspects of the private sector that do not fall into these areas of service. Technology growth and extreme weather have brought on the need for some more specialized versions of meteorological services.

10.6.1 Extreme Weather

One area of growing focus is that of damage estimation. In this field, meteorologists are often sought to explain the chances of certain extreme weather events occurring as well as typical weather events. The professional analysis of meteorologists in this area of need allows companies to better insure themselves through policy and preparedness. Meteorologists map out trends of how systems are and have been forming, such as typical tropical development in the Atlantic and Pacific during years of stronger El Niños, or the growing chances of extreme drought in Australia or the west coast of the United States. Meteorologists also help estimate storm surge potential along the coasts, and how certain weather conditions can lead to greater potential of human dangers that have yet to be seen. An example of a lack of this specialized type of private-sector meteorology was seen during the forecasting/trending of the possibility of a large and strong hurricane striking the New Orleans area and overrunning the levee system in place to protect the city. This, of course, did occur in 2005 with Hurricane Katrina's devastating results (Chapter 7, Section 7.2.2). A smaller and more localized example is the 100-year flood potential in Boulder, Colorado on the east side of the Rocky Mountains. It was here that there was always a potential for a strong long-lasting rainstorm that could cause mountain runoff resulting in the flooding of the normally small creeks. This mountain runoff funnels powerful walls of water emptying out directly into the heart of the city. While this was always a possibility, there was a definite lack of preparedness and understanding of the actual devastation that could occur should this event take place, which it actually did in 2013.

The sector of energy services often uses meteorologists to help predict and analyze the effects that weather may have on energy-providing units. Readiness for extreme weather can help energy companies keep their equipment running, employees safe, and production on schedule. Private-sector meteorologists can look for specific factors that may interrupt energy businesses or benefit them. For example, having a more accurate and specialized forecasting of weather elements such as extreme cold weather outbreaks or an outbreak such as the one seen in the winter of 2015 across the northern United States can help oil and heating companies to better prepare for an increase in demand of their resources. In terms of electric power, extreme heat is important to be accurately forecast, such as the extreme heat event in India and the Middle East in the summer of 2015. Extreme heat calls upon greater electricity use to cool households, businesses, and people, placing an extra burden on electricity providers and their machinery, often leading to failure. Knowing how weather relates to energy is a valuable resource to energy companies, and it is often the private-sector meteorologist who is responsible for providing that detailed and intimate information.

Understanding snowfall amounts for certain companies is imperative to ensure safety and production. Places such as ski resorts have some of the greatest requirement for this type of weather forecasting. Meteorologists are often consulted to describe the conditions expected for upcoming ski seasons, nowcasts for large blizzards and avalanche risk, and to forecast the type of snow expected based on temperatures and wind. Meteorologists are also needed to help consult on whether snow-making machines will be called for to produce the amount and type of snow ski resorts depend on. Weather-forecasting services in this area improve safety, and can increase the satisfaction of visitors to resorts and events. It can also greatly help predict the revenue resorts can expect to receive, as they are then able to better prepare for lulls or spikes in their money-producing seasons.

10.6.2 Weather Technology

Weather technology has grown across the board. As discussed in Chapter 6, Section 6.4, the invention of the iPhone and other cellphones capable of downloading weather applications has brought weather to the fingertips of the whole world. These applications are another form of private weather sector services. The businesses involved in creating some of the applications often have staffed meteorologists, or consult out to private companies in order to improve the application and to provide the best weather information to their users. Other technologies are weather-based software entities, which houses many of the governmental models and data all in one location, giving the forecasters all the necessary information in order to provide their own services to their own clients. In this example, the creator of the software is often part of a private weather company and the users of this software are other private weather companies. Private weather companies also create their own weather observation devices (radars) and often sell their services and products to governmental agencies, universities, and weather enthusiasts.

One final example of a technical weather service often developed and run by meteorologists is weather modification. The idea of weather modification dates back to ancient times, as discussed in Chapter 1, Section 1.7, with Native American and other folklore involving rain dancing rituals. From then on there have been numerous attempts not just to forecast weather but also to alter it. An example of a more scientific approach to early weather modification

is in 1890. US meteorologist James Pollard Espy attempted to use gunpowder explosions to generate the friction and nuclei needed to create rain (Lee, 2014). After numerous attempts similar to this, attention then turned to cloud seeding as a main way to alter the weather (Figure 10.1). Cloud seeding is mainly performed by aircraft or rockets that inject silver iodide and other substances and chemicals into the atmosphere to imitate ice nuclei. Since precipitation creation depends on the balance of ice nuclei in the atmosphere, cloud seeding helps increase the number of nuclei. This allows for greater snowflake production, resulting in clouds that would normally not produce precipitation to be packed with snowflakes that will eventually fall as precipitation that makes it to the ground. A similar use is sometimes utilized to disperse fog from nearby airports ("Weather Modification and Cloud Seeding Fact Sheet," 2008).

Figure 10.1 Photo of the 1966 Untied States Weather Bureau (today the National Weather Service) hurricane cloud seeding experiment aircraft and personnel. Project Storm Fury was one of the earlier regulated cloud seeding experiments. *Source*: Courtesy of NOAA Photo Library.

Since the 1980s, NCAR scientists have been carrying out research projects in a number of countries, including Argentina, Australia, Indonesia, Italy, Mali, Mexico, Saudi Arabia, South Africa, Thailand, and the United Arab Emirates ("Weather Modification and Cloud Seeding Fact Sheet," 2008). While results of cloud seeding and the atmospheric, climate, and pollution aspects of the process of cloud seeding and weather modification is often disputed and debated, it has become another branch of the private sector, mostly involved in precipitation enhancement and hail suppression. This controversial field is growing in the private sector, with private meteorologists and companies putting a lot of privatized money into technologies and resources in order to create an atmospherically safe and meteorologically productive technique to modify weather, relaying their own results to scientific resources and governmental agencies.

Section Summary

From 1980 through the early twenty-first century, the world has seen a tremendous technological expansion in all aspects of life, but especially in terms of atmospheric and meteorological sciences. The abundance of a desire of constant weather information helped create 24/7 weather television networks, lead stories on national news networks, Internet websites, and cellphone applications. Media coverage of weather disasters and events exploded, displaying never-before-seen videos and pictures of the weather conditions and the aftereffects of extreme weather. Governmental agencies had to keep up with the growth in technology, creating computer weather models that go a long way to helping meteorologists become as accurate as ever in predicting weather conditions and events. Radar and satellite technology allowed for views inside and above weather systems that previously were never conceived as being possible. Governmental agencies across the world expanded nationally as well as globally,

oftentimes working together to help protect and educate the masses and less-developed countries. The private sector in the field of meteorology is of great value and importance to many industries, companies, and the public across the world. Each country has a different way of going about how the commercialized weather business operates, some with complete detachment from the private sector, some with a stronger influence with revenue-making services. The overall objective is to end up furthering the field of meteorology and the services that it provides to the public. Technology has enabled services and fields of meteorology that were once considered to not have the possibility of success to become a necessity for governments, private companies, and the public around the world.

References

Book References

Aichholzer, G. and Burkert, H. (2004) *Public Sector Information in the Digital Age: Between Markets, Public Management and Citizens' Rights*, Edward Elgar Publishing, Cheltenham.

Committee on Partnerships in Weather and Climate Services, Committee on Geophysical and Environmental Data, and National Research Committee (2003) *Fair Weather: Effective Partnerships in Weather and Climate Services*, National Academic Press, Washington DC.

Cummings, R., Dinar, A., and Olson, D. (1996) *New Evaluation Procedures for a New Generation of Water-Related Projects*, International Bank for Reconstruction and Development/The World Bank, Washington DC.

Fleming, G. (2005) *Guidelines on Weather Broadcasting and the Use of Radio for the Delivery of Weather Information*, PWS-12, WMO/TD No. 1278, World Meteorological Organization.

House of Commons Select Committee on Science and Technology (2012) *Science in the Met Office: Thirteenth report of session 2010–12*, The Stationery Office, London.

JMBSC (2015) Japan Meteorological Business Support Center (JMBSC), Japan Meteorological Business Support Center, Tokyo, pp. 2–6.

World Bank (2008) *Weather and Climate Services in Europe and Central Asia: A Regional Review*, International Bank for Reconstruction and Development/The World Bank, Washington DC.

Journal/Report References

Gunasekera, D. (2004) Economic issues relating to meteorological services provision. *BMRC Research Report*, No. 102, August 2004, Bureau of Meteorology Research Centre, Australia.

Sarukhanian, E.I. and Walker, J.M. (n.d.) The International Meteorological Organization (IMO) 1879–1950, http://ane4bf-datap1.s3-eu-west-1.amazonaws.com/wmocms/s3fs-public/The_International_Meteorological_Organization_IMO_1879-1950.pdf?Td4kDN4kyR7yiUv5m1.y4vJTHhQ._STc. 14 January 2017.

Smith, M.R. (1999) The Future of the "Public-Private Partnership" Toward a More Synergistic Relationship in the 21st Century. *Weather Zine*, 19 (December).

Spiegler, D. (2007) The Private Sector in Meteorology: An Update. *Bulletin of the American Meteorological Society*, 88 (8), 1–20.

Website References

"About AMS." (2015) American Meteorological Society, AMS, https://www2.ametsoc.org/ams/index.cfm/about-ams, accessed 19 September 2015.

"About NCAR." (2015) The National Center for Atmospheric Research, National Science Foundation, https://ncar.ucar.edu/about-ncar, accessed 19 September 2015.

"About the National Weather Association." (2015) National Weather Association, NWA, http://www.nwas.org/about.php, accessed 19 September 2015.

"About Us." (n.d.) Royal Meteorological Society, RMetS, http://www.rmets.org/about-us, accessed 19 September 2015.

"An Introduction to the Meteorological Society of Japan." (2008) Meteorological Society of Japan, MSJ, http://www.metsoc.jp/E/intro-e.html, accessed 19 September 2015.

"Canadian Meteorological Private Sector." (2015) Environment Canada, Government of Canada, http://www.ec.gc.ca/meteo-weather/default.asp?lang=En&n=8DB56DCD-1, accessed 16 September 2015.

"Credibility of National Meteorological Services." (n.d.) Public Weather Services (PWS), World Meteorological Organization, https://www.wmo.int/pages/prog/amp/pwsp/visibilitystatusofnmss_en.htm, accessed 12 September 2015.

"Europe's Largest Private Weather Company Opens Office in Singapore." (2012) Press Corner. MeteoGroup, http://www.meteogroup.com/en/gb/about-us/press-corner/press/article/europes-largest-private-weather-company-opens-office-in-singapore.html, accessed 16 September 2015.

"Goals and Objectives." (2013) Board Goals and 2008 Objectives, AMS, https://www.ametsoc.org/boardpges/bpsm/goalsandobjectives.html, accessed 16 September 2015.

"History." (n.d.) About Us: Japan Meteorological Agency. JMA, http://www.jma.go.jp/jma/en/Background/history.html, accessed 19 September 2015.

Lee, J. (2014) The History of Cloud Seeding: From Pluviculture to Hurricane Hacking. Climate Viewer News LLC, http://climateviewer.com/2014/03/25/history-cloud-seeding-pluviculture-hurricane-hacking, accessed 16 September 2015.

"Organisations." (n.d.) Royal Meteorological Society, RMetS, http://www.rmets.org/our-activities/vocational-qualifications/organisations, accessed 16 September 2015.

"Policy on Partnerships in the Provision of Environmental Information." (n.d.) Partnership Policy, NOAA, http://www.noaa.gov/partnershippolicy, accessed 16 September 2015.

"Policy Statement on the Weather Service/Private Sector Roles." (1991) Industrial Meteorology,

NWS & NOAA, http://www.nws.noaa.gov/im/fedreg.htm, accessed 16 September 2015.

"Pricing Policy." (2014) Met Office, Met Office, http://www.metoffice.gov.uk/about-us/legal/pricing-policy, accessed 16 September 2015.

"Weather Modification and Cloud Seeding Fact Sheet." (2008) University Corporation for Atmospheric Research, NCAR & UCAR, https://www2.ucar.edu/news/weather-modification-and-cloud-seeding-fact-sheet, accessed 16 September 2015.

"What We Do." (n.d.) The United Nations Office for Disaster Risk Reduction, UNISDR, http://www.unisdr.org/we Accessed, accessed 13 September 2015.

Figure References – In Order of Appearance

"NOAA Photo Library." (n.d.) *NOAA*, http://www.photolib.noaa.gov/htmls/wea01153.htm. 18 August 2016.

Section IV

Current and Future Ideologies of Climate Change and Meteorological Processes

With the first half of this book surrounding the history of weather forecasting, we now enter into the issues of today and beyond. This section will discuss climate change and the various theories and evidence within it. Topics such as extreme weather and how climate change may be a causing factor of some of today's most recent atmospheric and meteorological activity will be discussed. Also, a discussion on the possible effects that solar activity has on weather, as well as the effects of weather on health and human activity will be presented, among other numerous topics. The one thing that they all have in common is that most, if not all, of the information provided for this text is based on scientific theory. Science is linked to evidence and data, and is derived from observations, measurements, calculations, and experiments. Scientists dispute, debate, and argue methods, accuracy, persuasion, and repeatability of data collection. Data collection in science cannot be derived from computer models, because they cannot virtually conform to earth's atmospheric, oceanic, and environmental processes. Science and scientific theory is never settled nor confirmed. To enable the progression of science, theories and hypotheses must be abandoned in order to create new explanations for the validation of evidence. Once evidence or a theory can be falsified, new progress can begin to develop, leading the way for further ideas and hypotheses to be explored to their furthest potential.

11

Climate Change

Nicole Gallicchio

Climatic discussions usually lead to a range of solutions and a great many conflicting opinions as to the cause of climatic change and what the effect of that change will or may be. Climate has and always will dictate humankind's actions and activities. Climate forecasting, unlike that of weather forecasting, is much more difficult. Despite all the advancements in the field, there are still extreme limitations on the accuracy of climate and long-range weather forecasting. No matter how powerful the world's greatest super-computer is, the calculations it is capable of performing still fall far short of being able to embrace all of the complexities of atmospheric-influenced processes that occur throughout earth, incorporating the earth's oceans and atmosphere, the sun, and the cosmos. Therefore, it can be hard to give the public a set of clear, reliable data with which to explain climate change and, moreover, suggest ways to combat it and protect our habitat. Many myths and misconceptions surround discussions of warming or cooling and cyclical changes of earth's atmosphere. And mixed messages and uncertainty serve only to confuse the public. When anyone researches climate change, be it in the library, online, by watching television programs or movies, or by reading scientific journals, etc., there is such an array of (often unsupported and conflicting) opinions, assertions, data, and predictions that it is impossible to know with any certainty in what way, why, or even whether

the climate is changing. When explaining climate change to society, it may be best to state that science is a way of observing, documenting, organizing, analyzing, interpreting, testing, and collecting data in order to logically explain and attempt to understand nature and human society. Hypotheses are a function of science that can increase our knowledge and understanding of the earth we live on. Testing and developing theories are the foundation of advancements. With earth ever changing, there will always be a need for science. Not all predictions and conclusions that will be drawn will be found to be accurate or correct; however, the beauty of evolution is that even in error we can find answers.

11.1 Climate Discussions

Concerns have been raised about the future of earth's climate. Throughout history earth's climate has changed, there is no denying that. Earth-orbiting satellites and other technological advances have enabled scientists to see the big picture, collecting many different types of information about our planet and its climate on a global scale. This body of data and evidence, collected over many years, reveals the signals of a changing climate ("Climate Change. How do we know?" 2016). This evidence shows that climatic conditions have changed in the past, and that nothing suggests that it will not

The Evolution of Meteorology: A Look into the Past, Present, and Future of Weather Forecasting, First Edition.
Kevin Anthony Teague and Nicole Gallicchio.
© 2017 John Wiley & Sons Ltd. Published 2017 by John Wiley & Sons Ltd.

continue to change in the future (Ahrens, 2007, p. 432). Scientists piece together available evidence in order to reproduce and examine climate cycles. Unfortunately, to our dismay, the available evidence for past climates can be less than perfect. It provides us with more of an overall generalization, which can sometimes be problematic. Fortunately, there are several different techniques for examining past climate data which enable a wide range of evidence to be pieced together to allow us to make global climatic generalizations.

There are many controversies surrounding climate change and its causes, the current situation, and the future of our planet. Our planet is 4.543 billion years old, and has gone through numerous climate changes, all in response to several different factors. Human involvement in earth's climate change is continuously being monitored and questioned. New technologies over the past 300 years have led to inventions and theories that were never believed to have been possible. When discussing climate change, the major topics that are focused on involve temperature, weather patterns, sea and ice levels, and the ozone. Throughout history, climate change attributions have been linked to very small variations in earth's orbit, which in turn changes the amount of solar energy our planet receives ("Climate Change: How do we know?" 2016). There have also been forces outside of earth that have enabled dramatic, drastic, and devastating climatic conditions, such as impacts from a meteor or an asteroid. During all the past changes to the earth's atmosphere, earth's surface has in turn undergone extensive modifications. Such modifications include plate tectonics, which is the theory that entails earth's outer shell being composed of large plates that fit together shifting and sliding around above a layer of molten rock. Further explanation and discussion upon the connection between climate change and plate tectonics can be seen in Chapter 13, Section 13.2.

Earth's climate is that of a complex, nonlinear system consisting of several smaller complex nonlinear subsystems ("Climate Basics," 2014).

Climate has always changed, and will always be changing. There would be an alarming amount of concern should earth's climate remain stagnant. Every part of earth's system is in its own way complex, and has distinct characteristics. For example, parts of this system include the cryosphere, the lithosphere, the oceans, the biosphere, and the atmosphere. Each of the subsystems is influenced by the other subsystems, and is driven by "external forcing." External forcing factors are processes external to earth and its atmosphere and include factors such as galactic variations, orbital variations, and sunspots. All these factors greatly influence earth's climate ("External Forcing Factors," n.d.). We need to explore every option we know about earth, its cycles, and its influences. Abandoning all possibilities by focusing on one factor, function, or variable in regards to earth's climate change leaves us close minded and vulnerable to even more potential unforeseen threats to existence.

Climate change has been hotly debated since the beginning of the Industrial Revolution. The increase in pollutants and chemicals entering our atmosphere that are derived from human expansion and technology raises numerous questions about how mankind is affecting the natural cycle of earth's climate. There is an extremely popular topic circling around the amount of carbon dioxide that is pumped into the atmosphere and its connection to human activities. There is much discussion and concern regarding carbon dioxide and methane entering the atmosphere through natural cycles started by a warming of earth, and the consequences of this. In order to get an understanding of what our climatic trends in the future will be, we need to know where our climate has been, where it is now, and what driving forces influence climate and weather.

Climate discussions should revolve around how human activities will affect earth's climate and what kind of weather and atmospheric conditions will surround a warming or cooling climate. These discussions will lead to questions about humankind's adaptability to survive either extreme warming or cooling, or an outside force, either of which can lead to extinctions and drastic biospherical changes. These are all viable discussions regarding the continuance of existence of earth's biological organisms.

11.2 Earth's Lifetime Fluctuations

When we look at earth's climate as a whole we find several fluctuations in atmospheric conditions. Earth has gone through cyclical changes, on all time scales, well before humans were roaming the planet. Earth enters warming and cooling periods, coupled with ice ages and interglacial periods. These changes are naturally caused by variations in sunlight, solar energy, volcanic activity, and changes in earth's orbit. Generally speaking, much more is known about earth's history in the past 500 million years than in the previous four billion years. When examining past climate changes, intervals of glaciation and atmospheric composition changes are always taken into consideration. Scientists also have to consider the interactions caused by the movement of earths plates. The causes of earth's fluctuations are not straightforward or consistent. Climate, weather, and plate tectonics will always be ever-changing sciences. Ever since the Pre-Cambrian period that occurred 600 million years ago, ice ages have occurred over widely spread intervals of geologic time, approximately 200 million years, lasting for millions to tens of millions of years (Maasch, 1997). An ice age is a long interval of time when global temperatures collaterally become relatively cold enough for glaciers to cover much of the earth. Conversely, an interglacial period is a long interval of time when a geological interval of warmer global average temperatures occurs. There have been at least five major ice ages throughout earth's history. The earliest ice age that we know about occurred two billion years ago.

The question of what can cause an ice age and interglacial period is a very complex one. As we know by now, several factors contribute to climate variations. Changes in oceanic and atmospheric circulation patterns, varying concentrations of atmospheric carbon dioxide, and volcanic eruptions are key factors that play into initiating ice ages. A significant initiator is that the tectonic plates are constantly moving. Plate motions lead to cycles of ocean basin growth and destruction. This movement is known as "Wilson cycles," which involves opening and closing of the ocean basins. The opening and closing of the basins involves rifting, convergence, plate collision, and mountain building. This plate movement causes continents to rearrange, redirect, and upset the ocean currents. The flow of ocean circulation patterns can restrict warm water flow from the equator to the poles, creating ice sheets to develop and set another ice age in motion (Eldredge and Biek, 2010).

The timing of the intervals between glacial and interglacial periods can be governed by a cycle change in earth's orbit. Earth's orbit affects the amount of sunlight that reaches the different parts of earth's surface due to orbital variations. Since 1920, it has been theorized that variations in earth's orbit are due to eccentricity, axial tilt, and precession. For the following explanations, it should be noted that the point at which the earth passes closest to the sun is called the "perihelion," and the furthest point from the sun is called the "aphelion." Also, the term "precession" means the change in orientation of the rotational axis of our rotating planet (wobbling on its axis). The amplification or dampening of seasonal climatic variability, due to the precession of the equinoxes (perihelion and aphelion), is caused by the precessional cycle. It does not affect the total amount of solar energy that the earth receives, but only the hemispheric distribution over time ("Precession," n.d.). For example, if it is summer in the northern hemisphere and earth is in a perihelion phase then that summer is likely to have more extremes. Conversely, if the northern hemisphere experiences an aphelion during the summer then the season is more likely to be less severe. The cycle fluctuates roughly on a 22,000- to 26,000-year basis. The National Climate Data Center (NCDC) of

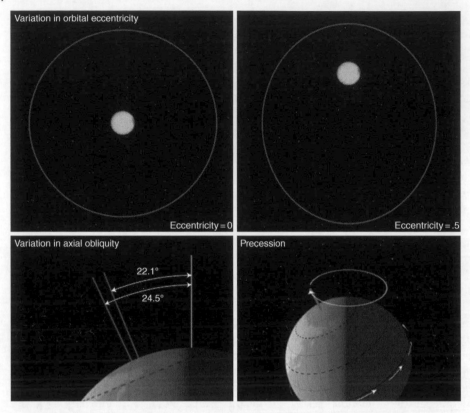

Figure 11.1 The earth goes through cyclical changes while it orbits the sun. These cycles include the changes in earth's precession due to wobbles on its axis, changes in tilt of the axis, and changes in the eccentricity of the orbit. All changes affect the weather and seasons seen at earth's surface. *Source*: NASA.

the National Oceanic and Atmospheric Administration (NOAA) states that the earth is currently in a period of the cycle that brings perihelion during northern hemisphere winters (Andrews, 2008) (Figure 11.1).

"Eccentricity" is defined as the measure of deviation of an orbit from its circularity. To explain further, a perfectly circular orbit would have an eccentricity of zero; the higher the eccentricity, the more elliptical an orbit is. Earth's orbital eccentricity varies over time, and takes approximately 100,000 years for earth to undergo a full cycle. In a period of high eccentricity, of a more elliptical orbit, radiation exposure from the sun can fluctuate at a higher probability between periods of aphelion and perihelion. Conversely, it is reversed during periods of lower eccentricity. Earth's current

eccentricity is nearly at its minimum in the cycle, making our orbit more circular ("Milankovitch Cycles and Glaciation," n.d.). This raises questions and comparisons between climate effects of a lower eccentricity and solar radiation exposure. Theory would coincide with the idea that low eccentricity equates to lower solar exposure, leading to ultimately less-than or equal-to average temperatures due to radiation exposure in respect to eccentricity.

The tilt of the earth's axis is most notably known for controlling the seasons in the northern and southern hemispheres. When looking on a large scale the angle at which the earth tilts has a variation that is in accordance with a 40,000-year cycle. Earth's tilt varies between 22.1 and 24.5 degrees. When the degree of tilt is higher, the seasons on earth can be more severe, and when the degree is lower the seasons are likely to be less severe. Currently, earth is in the middle of a decreasing phase at obliquity of 23.5 degrees ("Milutin Milankovitch (1897–1958)," n.d.). When relating to climate change on earth, this portion of astronomical effects would place us in a phase that is less likely to experience drastic seasonal changes.

Past ice ages can be reconstructed by piecing together derived data from paleoclimate studies. An example of earth's fluctuations can be seen as earth was emerging out of the last glacial cycle. Approximately 12,800 years ago, temperatures dropped dramatically for several decades. This coincided with a layer of material that is thought to have been part of a major cosmic impact, which in turn blocked the sunlight, causing a significant and rapid cooling down. About 1300 years later, temperatures locally spiked as much as 20°F within several years. Sudden fluctuations like this have occurred at least 24 times during the past 100,000 years (Eldredge and Biek, 2010). It is uncertain if we can expect temperature fluctuations to increase or decrease, and whether they will be corrected naturally over time, as other cyclical cycles have.

The period during which the last ice age occurred is known as the Pleistocene epoch. An epoch can be defined as a span of time in which a new beginning occurs. During the Pleistocene epoch, Homo sapiens evolved, and could be found almost all over the planet by the end of this period, approximately 11,700 years ago (Zimmermann, 2013). The earth is currently in an interglacial period, the Holocene epoch, which began 12,000–11,500 years ago at the close of the Paleolithic Ice Age (Bagley, 2013). When the Paleolithic Ice Age came to a close, climate changed from extreme cold to a warming trend. This trend gave way to forests, and extinction of animals who had adapted to cold, such as mammoths and wooly rhinoceros. Humans were able to adapt to the warmer climate, including changing their diet and agricultural ways. During the evolution of earth, there have been at least five major mass extinction events. A mass extinction is a rapid and widespread decrease in the

amount of life on earth. Many questions have arisen regarding whether earth is in the midst of approaching its next mass extinction. Will the next mass extinction come from human activity, natural causes, or not-of-this-world effects? At the moment, habitat destruction is the leading cause of species extinction today (Bagley, 2013). The predictions of an approaching mass extinction stem from the climate change that evidently is rapidly occurring, whether this be from human activity, natural effects, or both. "In the past, global climate change has often been synchronous with mass extinction" (Bagley, 2013). It is argued in several papers that man has already begun to enter a new epoch: the Anthropocene epoch. In 2002, Paul Crutzen, a Nobel Prize winner, coined the term in order to reflect the changes humans have brought upon the world since the Industrial Revolution (Owen, 2010). The argument goes that the change was already happening before the Industrial Revolution took place, while the rebuttal to this is that the rate of change has significantly increased due to human activity. Some scientists even argue that we have not entered the Holocene epoch, however, and are still in an ice age that should go on for another million years, oscillating between cold and warm periods (Owen, 2010). Earth still contains ice caps at its poles, supporting the argument that earth is simply oscillating between warming and cooling periods during a continued ice age.

11.3   Past Climate Data

When comparing the atmospheric, meteorological, and climatological data that exist to the age of earth, it seems like quite a small amount of data. How do we, as humans, determine past climates if we were not there to record it? Scientists use numerous techniques all around the globe to piece evidence together. Some of these techniques include using ice cores, fossil pollen, ocean and lake sediments, loess, glaciers, speleothems, trees, boreholes, and meteorological records. The data collected by these techniques allow for the study of a past climate, also known as "paleoclimate." Scientists use imprints in different geophysical, environmental, and biological formations, more scientifically known as "proxies," to examine past climates. The World Data Service for Paleoclimatology archives a variety of paleoclimate proxy data and climate reconstructions from all over the globe ("World Data Centers," n.d.). Proxy data are contributed to from the efforts of scientists around the world, publishing the results of their research and data.

11.3.1 Coring

Arguably the most important, resourceful, and effective form of collecting past atmospheric and climatic data is ice coring. Ice cores can be used by reading oxygen isotopes and air bubbles to determine the temperature at the time the ice was deposited and the measure of carbon dioxide and methane concentrations at the time the bubbles were formed. They are cylindrical samples of ice drilled from a glacier or continent. This same technique is used on the ocean floor to study ocean sediments.

Ocean sediment studies have allowed scientists to plot a climatological chronology and details regarding environmental change. Sediments made up of mineral grains from continents can provide information regarding ocean currents. When sediments are dumped in the ocean, either by rivers or by wind, it is left to settle on the ocean floor. The distribution of these sediments can reveal how strong or weak currents were and in which direction the flow was. This method of current flow can be seen by the way icebergs move and how sediments sink to the ocean floor, along with gaining information for the point that the ocean water was warm enough to melt the ice. These sediments are studied through ocean coring. Ocean cores provide a record of preserved sediment with a water interface that sits on top of the sediment surface.

11.3.2 Varve

"Varve" is defined as an annual sediment layer. For the purposes of examining past climate,

varve is referred to as the separate components of annual layers in glacial lake sediments. Lake and ocean sediments provide insight into environmental conditions and water temperature by examining such things as pollen in sediments and primitive shelled animals ("How Do We Determine Past Climate?" 2016). The varying thicknesses of the varve is the key to providing information about climatic conditions. Thick lake varve results from a warmer temperature, allowing for an increase in melting and therefore more deposition of sediments. The contrary occurs from a colder environment, with thinner varve resulting from less sediment depositing from a more solid glacier.

11.3.3 Loess

Loess, an extensive semi-compact yellow/gray deposition of typically wind-borne sediments, can provide informative details on past wind and moisture content of a region. Even though loess is mostly created by wind, it can also be formed by glaciers. This occurs when glaciers grind rocks down to a fine powder. Streams will carry the fine powder to the end of the glacier, and loess will form. The relationships between proxies and sedimentation rates are used for climatic interpretations based on different age models (Stevens, *et al.*, 2007). The implications of loess allow for the reconstruction of a climate and comparisons to be made between paleoclimate records and evidence from ice and ocean cores.

11.3.4 Glacial Deposits

Glaciers come in various sizes. They are slowly moving masses or rivers of ice formed by the accumulation and compaction of snow on mountains or near the poles. Glacial deposits of rock and debris can allow the location and presence of past glacial sizes and location to be inferred. Speleothems are mineral deposits that are formed within underground caverns. Some of these mineral deposits such as stalagmites and stalactites can be radiometrically dated using glacial deposits embedded within speleothems.

Scientists can date the layers of speleothem by measuring the amounts of uranium that has decayed. Over time, uranium predictably turns into thorium, a radioactive element; therefore, scientists are able to tell how old a layer of rock is by finding the ratio of uranium to thorium (Riebeek and Simmon, 2005).

11.3.5 Dendrochronology

Another common form of studying past climates is the study of tree rings: dendrochronology. Trees contain rings that are proportional to the soil moisture, temperature, and growing conditions. Tree rings are the number of concentric rings in the cross-section of a tree trunk, representing a single year's growth. Annual rings in temperate forest can be used to piece together the climate of the past. A single tree ring can only give limited information about the environmental conditions; however, when hundreds of thousands of tree-ring records are combined, they can tell us much more. Even though this type of paleoclimatology has its limitations, its reliability has grown and is used across multiple disciplines. There are extensive and comprehensive data sets throughout Europe and North America that aid in the study of paleoclimatology and prehistory (Mason, 2016).

11.3.6 Boreholes

Boreholes are deep narrow holes made in the ground, commonly used to locate water and oil, and can be sometimes used to conclude past surface temperatures. This method allows several hundred years of temperature variations to be studied. This is done by using a piece of "earth" in which the ground is not disturbed by groundwater flow. "Soil temperature anomalies slowly propagate into the ground, depending on the heat conductivity and thermal capacity of the layers. Under suitable conditions the geological factors affecting the geothermal gradient can be isolated, so the history of past surface temperatures can be inferred from small temperature anomalies along the length of a borehole" (Linacre and Geerts, 1998).

11.3.7 Climate Data Limitations

It is hard to be certain that all the data predictions from all the different climatological "aging" techniques are accurate and align properly with the past. Each of these paleoclimatology techniques, as well as others not mentioned, have weaknesses resulting in limitations to research. Problems with reconstructing a past climate include accuracy, precision, chronology, and integration. One problem lies closely with that of sampling precision. In paleoclimate studies, close interval sampling is necessary to detect small changes in the proxies; however, this doesn't allow one to see changes on a large scale. These large-scale data sets are necessary to see significant changes in climate. However, taking large sampling intervals allows for a covering of small perturbations in data, without their skewing results. There is also a problem with the integration of paleoclimate and archeological data. In areas of good paleoclimate data, there is no sampling for archeological data to be measured against. In order to achieve better accuracy matches, different types of scientific research and data using different techniques should be linked, solidifying findings and past climate. Limitations also exist when studying tree ring patterns. Factors such as the timing of the growing season, the tree's placement on earth, and the length of the life of the tree all introduce additional variables to the study ("Topics in Studying Climate Change," n.d.). Ice coring provides an uncertain timeline even though the depth of an ice core is directly related to the passage of time. The magnitude of error and uncertainty is dependent upon the degree of ambiguity in identifying seasonal markers, and the likelihood of missing layers. Both of these configurations are functions of snow accumulation rate and location from which the sample is taken (Linacre and Geerts, 1998). Sampling uncertainty derived from noncontinuous markers, along with a degree of spatial uncertainty and physical relationship variables, is also a factor behind the margin of error in paleoclimate data.

It has become abundantly clear through proxy data and paleoclimatology that the rate and speed at which the climate can change can occur within mere decades. In order to understand these transitions, the reconstruction of past climates over a wide variety of temporal and geographical scales is necessary. The reconstruction becomes stronger once a multitude of fields are merged, using information from physics, geology, anthropology, archeology, history, chemistry, and oceanic and atmospheric sciences. An integration of proxy data from multiple outlets reduces the margin of uncertainty and generates more reliability when reconstructing a past climate.

11.4 Permafrost and Climate Change

After studying the past, it is then necessary to study the present and future. Should the climate be warming, then significant attention needs to be addressed to the different elements on earth that are and will be affected. Certain compositions on earth are greatly affected by climate change. When these compositions are affected, the result that comes only adds to the growth of climate change, which then only further alters the climate and earth, thus creating a vicious and intense cycle. Permafrost is a perfect example of this continuous cycle. Permafrost is a thick permanently frozen sublayer of soil, rock, or sediment. The classification of permafrost is solely based upon temperature, and does not take into consideration moisture. This layer must remain frozen for two or more years in order to be classified as permafrost. Although that classification of permafrost relies on temperature, moisture can aid the chemical composition and formation of the ground and its abilities to be classified as permafrost. Most permafrost is located in the high latitudes; however, some can be in the lower latitudes. The strength, characteristics and thickness of permafrost is dependent on the moisture and variations in air and ground temperature, as well as geothermal heating.

The depth in turn fluctuates depending on the seasonal cycles (Rafferty, 2011, p. 30). The upper surface of permafrost is called the "permafrost table." In permafrost areas the surface layer of the ground that freezes during the winter and thaws in the summer is called the "active layer." It is estimated that permafrost underlies 20% of earth's land surface. Permafrost is a result of earth's climate. It exists when the mean annual air temperature is 0 °C or colder. Such regions' atmospheric conditions would produce weather conditions such as being relatively dry with only little or short snowfalls, along with cold, long winters, and short, cool summers. The thickness of permafrost can fluctuate in two different regions with the same mean annual air temperature due to the conductivity of the ground and the geothermal gradient (Pewe, 2016). Figure 11.2 shows the spatially distributed modeling of permafrost throughout northern Alaska.

Solar radiation, vegetation, and snow cover affect the presence of permafrost. The main effect of vegetation in local permafrost areas is to protect and insulate the frozen ground from solar energy. When the vegetation becomes absent, due to natural processes, or by human disruption, the underlying permafrost becomes exposed and causes thawing. This thawing can either lower the permafrost table or may defrost it completely. Another insulator for permafrost is snow cover. Snow cover influences heat flow between the ground and atmosphere, affecting the distribution of permafrost (Rafferty, 2011, p. 36). The ratio of vegetation and snowfall are critical for the formation and existence of perennially frozen ground in certain areas of the world, depending on the extent and type of permafrost present.

Ice content of permafrost is a feature that affects human life. Ice in permafrost exists in various sizes and shapes, having definite distribution characteristics. Ground ice is grouped into five main categories: pore, segregated, foliated, pingo, and buried ice. Each category defines ice observed with different characteristics and formations. "World estimates of the

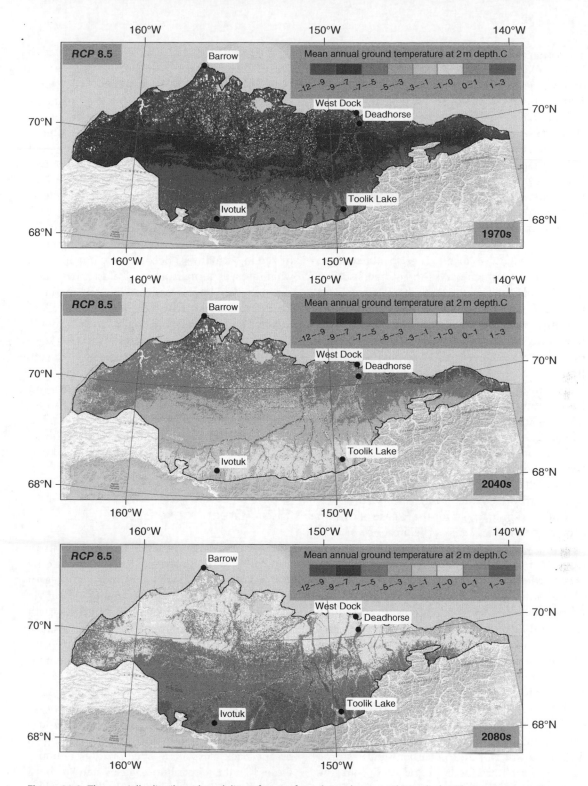

Figure 11.2 The spatially distributed modeling of permafrost throughout northern Alaska. The top image is the temperatures present in the 1970s. The bottom image is an estimate of the predicted temperature in the 2080s. The most drastic observation one can take from this comparison is the all-round general warming trend in all areas. Looking at the future predictions, temperatures are estimated to be above freezing for nearly all of northern Alaska, with the coldest locations to the far north at only around –1 to –3 °C. This is compared to the 1970s, when the warmest locations were only as high as –3 to –5 °C and as cold as –12 °C to the far north. *Source*: Courtesy of Dr. Dmitry Nicolsky. (*See color plate section for the color representation of this figure.*)

amount of ice in permafrost vary from 200,00 to 500,00 cubic km (49,000 to 122,000 cubic miles) or less than 1% of the total volume of the Earth" (Rafferty, 2011, p. 38).

Permafrost contains a lot of frozen organic matter. This organic matter consists of remains from organisms such as plants and animals and their waste products that have been frozen deep in the ground for thousands of years. This organic matter contains a lot of carbon as a result of the processes that took place during the last ice age, when great ice sheets covered most of the continents. As the ice sheets spread and shrunk, the rock beneath them was ground up into loess, or glacial flour, which was deposited onto the soil. These layers of soil built up over time, creating a thick layer. The active layer on top thaws each year allowing plants to grow; however, the roots and organic matter were consistently frozen into the permafrost where they cannot decay. As long as the permafrost remains frozen, there is no concern of any of this organic matter being released into the atmosphere. However, once the permafrost thaws, the organic matter will decay by being consumed and digested by microbes, which will release carbon dioxide or methane as a product. The determining factor for the release of carbon dioxide or methane gas is the presence of oxygen. If oxygen is present, the microbes will produce carbon dioxide; if not, methane will be the result. Methane gas is naturally in the atmosphere, with the chemical composition of one carbon and four hydrogen atoms. Even though it is naturally found in the atmosphere, it traps heat approximately 20 times as efficiently as carbon dioxide and would cause an increase in atmospheric temperature even more so than carbon dioxide does from the burning of fossil fuels ("All About Frozen Ground," 2016).

The types of regions that would produce methane instead of carbon dioxide are swamps and wetlands. Not all the carbon that is stored in the permafrost will be released into the atmosphere. The concern is that if a significant amount of methane and carbon dioxide is released into the atmosphere from thawing

permafrost then the atmosphere will warm significantly and rapidly. There are estimates of how much carbon is stored in permafrost currently that could be released into the atmosphere; however, new studies are being conducted to try to predict how fast the frozen carbon will decay, and how much methane and/or carbon dioxide will be released into the atmosphere. In relation to climate change, there are several questions that need to be scientifically addressed. First, how much carbon is currently frozen in permafrost? Second, how much and at what rate will permafrost thaw? Last, how much carbon would be released and in what chemical form? The answers to these questions will allow humans to determine and configure the modifications that need to be taken in order to keep earth's atmospheric temperature at a steady number. If methane and carbon dioxide released by permafrost thawing are found to increase atmospheric temperatures significantly, and no action is taken to offset this increase, then it will have a cyclical effect on earth's atmospheric, oceanic, and meteorological processes and elements ("All About Frozen Ground," 2016).

11.5 Greenhouse Gases

Gases within the atmosphere have the ability to absorb, hold, and store heat. This is not the only method or process that occurs in order for earth to stay warm; however, this trapping of heat via gases is an important role in what enables earth to sustain life and warmth. Many chemical compounds in earth's atmosphere can behave as greenhouse gases. These gases allow sunlight (relative shortwave energy) to reach the earth's surface unimpeded. As the shortwave energy heats the surface, longwave (infrared) energy is reradiated to the atmosphere. The greenhouse gases absorb this energy and allow less energy to escape back to space ("What Are Greenhouse Gases?" n.d.). Greenhouse gases can be both naturally occurring and synthetic. The two most common greenhouse gases are water vapor and carbon dioxide. Other greenhouse gases include nitrous oxide, methane, and synthetic gases such as chlorofluorocarbons (CFCs), perfluorocarbons (PFCs), hydrofluorocarbons (HFCs), and sulfur hexafluoride (SF6). Methane is an important greenhouse gas due to its fast and powerful warming effect, even though, relatively speaking, there is not a large quantity of it in the atmosphere. As explained in Section 11.4, the storage of these gases in permafrost is of concern due to the melting and potential release of a significant amount of carbon dioxide and methane gas. Research stations in the Antarctic have studied ice cores in order to measure the amount of greenhouse gases in the atmosphere in the past. These ice core studies have found an increase in carbon dioxide and methane gas, linking the increase to a rapidly warming atmosphere. As technology has evolved and global populations have increased the reliance on fossil fuels has dramatically increased the emissions of these greenhouse gases.

11.6 The Intergovernmental Panel on Climate Change

With more information being gathered across the world on the past, current, and future state of the climate of earth, it has become necessary to debate the issues (concerns, research, data, theories, etc.) in a constructive way and with a global perspective. In order to bring the greatest minds together, the United Nations Environment Programme and the World Meteorological Organization formed the Intergovernmental Panel on Climate Change (IPCC) in 1988. The IPCC was established with the intention of providing the world with a clear scientific view of the current state of knowledge in climate change, and its potential environmental and socio-economic impacts ("Organization," n.d.). The IPCC reviews and assesses the most recent scientific information and research provided worldwide relevant to climate change. It does not conduct any research, and nor does it monitor climate-related data or parameters. The assessments are written by leading scientists

who volunteer their time and expertise. These assessment reports are released every few years and undergo numerous rounds of drafting and review to ensure they are comprehensive and objective ("IPCC Factsheet: What is the IPCC?" 2013). Recent assessments include the AR4, the Copenhagen Diagnosis, and the AR5. These reports cover the scientific, technical, and socio-economic information that is relative to human-induced climate change. It explains the risks and impacts that are associated with climate change. "The main goal of these assessments is to inform international policy and negotiations on climate-related issues." ("The IPCC: Who Are They and Why Do Their Climate Reports Matter?" n.d.). The first assessments that were prepared supported the establishment of the United Nations Framework Convention on Climate Change (UNFCCC), an organization with the objective "to stabilize greenhouse gas concentrations in the atmosphere at a level that will dangerous human interference with the climate system" ("UN Climate Change Newsroom," n.d.).

There is some scientific unrest and disagreement with the results and purpose of the IPCC and its reports. Some question the motives behind the reports, making claims that the documents are designated specifically to fund politics and do not resemble current trending atmospheric conditions. Many have questioned whether the IPCC is misrepresenting the scientific global community. There can be a large gap between scientific reporting and society's understanding of the statements and hypotheses being expressed. Questions arise about the manipulation of data and the practice of picking data that are convenient and extrapolating their theories. Another topic that typically is discoursed is the lack of modeling that can be done to prove the reports' conclusions. Modeling that is conducted could also exaggerate or project wrong data, causing extreme uncertainty, possibly even overestimating climate change and the dangers it poses to humans. The contrary can also be questioned, if the IPCC is underestimating not only climate change but also its rate, and the impact that human activity is having upon it. These situations only enhance the gap between the scientific community and society, which reduces society's confidence in any proclamations made and thus lessens how seriously it takes any (potentially dire) warnings.

11.6.1 The Copenhagen Diagnosis

The Copenhagen Diagnosis is a follow-up report to the previous IPCC's Working Group 1 Report. The report serves as a midpoint between the IPCC Assessment Report AR4 and the IPCC AR5. It was published in 2009 and covers numerous components of the current climate and its ever-changing cycle. More specifically, the report goes into details covering greenhouse emissions, atmospheric concentrations, and the analysis of the global carbon cycle. It includes all major components of the cryosphere, atmosphere, and the land and surface. Paleoclimate, extreme events, sea levels, future projections, abrupt changes, and tipping points are also included in the report.

The report begins by discussing greenhouse gas emissions such as carbon dioxide and methane gas. In 2008, scientists found that global carbon dioxide emissions from fossil fuel burning in 2008 were 40% higher than those in 1990, with a threefold acceleration over the span from 1990 to 2008. They also recorded that global carbon dioxide emissions from the burning of fossil fuels were tracking near the highest scenarios considered by the IPPC. Lastly, the report states that the fraction of carbon dioxide emissions absorbed by the land and ocean carbon dioxide reservoirs likely decreased by 5% from the 1950s (Allison, *et al.*, 2009, p. 8).

Global atmospheric temperatures are always a topic of discussion. The Copenhagen Diagnosis reports that atmospheric warming trends continued to climb despite 2008 being cooler than 2007. The report discusses that every year from 2001 to 2008 had been among the top 10 warmest since instrumental recording began. Effects of solar radiation along with natural and human-induced warming are topics of the research conducted. The studies

recorded that the sun contributed about 10% of surface warming in the last century. The Copenhagen Diagnosis reports that incoming solar radiation has been nearly constant over the past 50 years, apart from the 11-year solar cycle. Other natural factors, such as volcanic eruptions and El Niño events, have only contributed to short-term temperature variations. There have been numerous questions regarding the atmosphere containing more water vapor, due to a warming climate. The report indicates that, yes, water vapor becomes more plentiful in a warmer atmosphere. This has been noticed from satellite data, showing that from 1998 to 2008 the atmospheric moisture content increased (Allison, *et al.*, 2009, p. 12).

Extreme events and land-surface interactions are documented in the report describing the impacts from climate change. The report indicates that increases in hot extremes and decreases in cold extremes are expected to amplify in the future, along with increases in precipitation extremes. These findings are of great importance due to their quick impacts on society, the environment, and ecosystems (Allison, *et al.*, 2009, p. 15). Vegetation is also strongly affected by atmospheric pollutants. Land cover change, especially deforestation, has a large impact on the carbon dioxide process. The report observes that observations through the 2005 drought in Amazonia suggested that the tropical forests could become a strong carbon source if precipitation amounts decline in the future. The Copenhagen Diagnosis suggests that avoiding tropical deforestation could lead to a prevention of up to 20% of human-induced carbon dioxide emissions, helping to maintain biodiversity (Allison, *et al.*, 2009, p. 19).

As discussed in Section 11.4, the changes in earth's permafrost is an avid question in regards to climate change. The Copenhagen Diagnosis refers to the large potential sources of methane and carbon dioxide released into the atmosphere with the thawing of permafrost, especially in the northern hemisphere. The report also discusses the significant source of methane that exists as hydrates beneath the deep ocean floor.

The IPCC states that it is unlikely that the large release of these gases will occur this century (Allison, *et al.*, 2009, p. 21). However, if any amount of these gases was released, it would amplify the warming of our atmosphere.

Global ice, in its many forms on earth, will always be considered, for as long as climate change remains an issue. According to the Copenhagen Diagnosis, there is evidence of increased melting of glaciers and ice caps since the mid-1990s. In turn with ice melting, the rise of global sea levels is expected. Recorded in the report, global sea levels have increased from a rate of 0.8 mm (0.031 in) per year in the 1990s to 1.2 mm (0.047 in) per year in 2008 (Allison, *et al.*, 2009, p. 23). It was noted that the surface area of ice sheets of Greenland and Antarctica has seen a loss in ice mass at an accelerated rate since the mid-1990s, with both Greenland and Antarctica nearly equally contributing to sea-level rise (Allison, *et al.*, p. 24). The most significant factor in accelerated ice discharge from both land masses was documented to be the un-grounding of the glacier front from its bed, due to submarine ice melting (Allison, *et al.*, 2009, p. 26). Ice shelves are floating sheets of ice that are attached to the coastline. Ice shelves are of considerable thickness, and mainly composed of ice that has flowed from the interior ice sheet, or that has been deposited as local snowfall (Allison, *et al.*, 2009, p. 27). The Copenhagen Diagnosis suggests destabilization of ice shelves along the Antarctic Peninsula. This is noted to be the result of a warming climate influencing ocean warming, producing ice sheet instability and mass loss.

The Copenhagen Diagnosis includes tipping points and abrupt changes that can occur due to human activity, leading to an irreversible change. It suggests transitions that earth could undergo, along with the greatest concerns that have the largest negative impacts. The closing sections of the report include lessons that we should be learning from the past as we attempt to reconstruct past climates. These reconstructions are done with using such techniques as ice coring and tree-ring readings.

Climate predictions were made in the IPCC AR4 report, and no new coordinated set of future climate model projections were undertaken between the AR4 report and the Copenhagen Diagnosis. The research being conducted at the time the Copenhagen Diagnosis was written was in preparation for the next round of IPCC simulations, for the Fifth Assessment Report (AR5). The AR5 was published in 2013 and expands on earth's current climate.

11.6.2 The Fifth Assessment Report (AR5) Overview

The Fifth Assessment Report (AR5) is published and explained in *Climate Change 2014*. This comprehensive assessment of the physical aspects of climate change focuses on the elements relevant to understanding past and future climate change. The AR5 includes findings in the following working group contributions: *The Physical Science Basis; Impacts, Adaptation, and Vulnerability* and *Mitigation of Climate Change*. The Fifth Assessment report was also a solidification of the AR4 and Copenhagen Diagnosis, along with drawing findings of the reports, *Renewable Energy Sources and Climate Change Mitigation*, and *Managing the Risks of Extreme Events and Disasters to Advance Climate Change Adaptations*. The AR5 found that the IPCC strengthened its earlier 2007 qualification of "likely" to "extremely likely that human influence has been the dominant cause of the observed warming since the mid-20th century" (Lovejoy, 2014). "This SYR [Synthesis Report] includes a consistent evaluation and assessment of uncertainties and risks; integrated costing and economic analysis; regional aspects; changes, impacts and responses related to water and earth systems, the carbon cycle including ocean acidification, cryosphere and sea level rise; as well as treatment of mitigation and adaptation options within the framework of sustainable development" (IPCC, 2014, p vii). The report also discusses aspects of climate change in relation to its impacts on all the systems on the planet. There are four major topics in the SYR report: Topic 1: Observed Changes and their Causes; Topic 2: Future Climate Changes Risks and Impacts; Topic 3: Future Pathways for Adaptation, Mitigation and Sustainable Development; and Topic 4: Adaptation and Mitigation.

Topic 1 focuses on human influence on the climate system and observational evidence of climate change with impacts caused by human contributions. The topic discusses observed changes and external influences on climate and differentiating the forces of climate change origins. It brings attention to the attribute's impacts on human natural systems to climate change. The changing probability of extreme events and their causes are also reported, along with risk context and adaptation and mitigation experience. The findings of the SYR indicate that "each of the last three decades has been successively warmer at the earth's surface than any preceding decade since 1850" (IPCC, 2014, p. 40). It also notes that ocean warming dominates the increase in energy stored in the climate system, and that it is very likely that regions of high surface salinity where evaporation is dominant have become more freshwater since the 1950s. Changes in evaporation and precipitation over the oceans lead to changes in the global water cycle; however, the report notes it only has medium confidence in this and that there is no evidence of a long-term trend. The report also discusses the acidification of the ocean, due to the uptake of CO_2. On the topic of extreme events, the SYR finds that extreme weather and climate events have been observed since about 1950, with some of these climate changes linked to human activity. These climate changes include decreases in cold temperature extremes, an increase in warm temperature extremes, an increase in extreme sea level highs, and an increase in the amount of heavy precipitations events. Portions of Topic 1 focus on the attribution of climate change, stating that, "it is extremely likely that more than half of the observed increase in global average surface temperatures from 1951 to 2010 was caused by

the anthropogenic increase in greenhouse gases concentrations and other anthropogenic forcings together" (IPCC, 2014, p. 48). The report also finds that there is a very high confidence in the seasonal activities and migrations patterns in many species of all classifications. This is attributed to different climate change aspects such as ecosystem disturbances, oxygen concentrations, water salinity changes, and temperature changes.

Topic 2 discusses future climate changes, risks, and impacts. It assesses projections of future climate change and resulting risks and impacts (IPCC, 2014, p. 56). The beginning of this section of the AR5 discusses key drivers of future climate and the basis for these projections. It goes into detail about CO_2 emissions, greenhouse gases, and the climate models and their uncertainty. There are projections that the surface temperatures will rise over all possible emission scenarios throughout the twenty-first century with it being "very likely" that heatwaves will be stronger and last longer, and that ocean waters will warm as well as acidify, all while continuing to increase in sea level heights (IPCC, 2014, p. 58). When looking at the future risks and impacts, the report states that those who will likely be most affected by existing risks and new risks will be those in disadvantaged communities and countries (IPCC, 2014, p. 64). It is stated that many land and marine life species face increased risks of extinction beyond the twenty-first century, and that redistribution of some species may cause issues with fisheries and other ecosystem services. Also, the threat of water scarcity as an increasing threat for parts of the world will become more likely (IPCC, 2014, p. 67).

Topic 3 is based on the future pathways, mitigation, and sustainable development. "Adaptation and mitigation are complementary strategies for reducing and managing the risks of climate change. Substantial emissions reductions over the next few decades can reduce climate risks in the 21st century and beyond, increase prospects for effective adaptation, reduce the costs and challenges of mitigation in the longer term and contribute to climate-resilient pathways for sustainable development" (IPCC, 2014, p. 76). Research suggests that making substantial cuts in greenhouse gases over the next few decades will have an enormous impact, limiting warming of earth over the latter half of the century. Without mitigation, irreversible impacts at the global level are highly likely (IPCC, 2014, p. 77). The common theme throughout this topic strongly states such scenarios where adaptation to the new climatic ways is at war with mitigation options. "There are numerous strategies and research to take on the benefits of both approaches, as well as the adverse side effects and risks that both options provide" (IPCC, 2014, p. 91).

Topic 4 of the AR5 involves the more detailed adaptation and mitigation options. The main point in this topic is the idea that, yes, there are many options out there, but not one single option will be able to solve the climate threat by itself. There has to be abundant cooperation across all scales. "Adaptation and mitigation responses are underpinned by common enabling factors. These include effective institutions and governance, innovation and investments in environmentally sound technologies and infrastructure, sustainable livelihoods and behavioral and lifestyle choices" (IPCC, 2014, p. 94). Some examples of areas discussed to manage risks of climate change include human development, by increasing and improving education, health facilities, and energy; disaster risk management, by increasing and improving early-warning systems and hazard mapping; and with spatial or land use planning, by improving zoning laws and managing the development in flood-prone areas (IPCC, 2014, p. 96). Some theories in the report on the adaption side are examples such as genetically modifying foods, sea walls, water trading, and habitat protection for vulnerable species (IPCC, 2014, p. 98). In order for any adaptation process or mitigation process to work, there has to be global cooperation involved. The UNFCCC is one of the main international forums battling all these issues involving climate change (IPCC, 2014, p. 102). There are a number of newer institutions setting out to tackle the adaptation option, but this option has generally received less attention than the options presented through mitigation, but economic instruments, such as loans and payments for environmental services, can help provide incentives for the anticipating and reduction of climate change impacts (IPCC, 2014, pp. 105, 107). Using both approaches (adaptation and mitigation) can end up creating synergies, which may attract more support and advance societal goals (IPCC, 2014, p. 112).

With 95% certainty that humans are the main cause of current global warming, the AR5 set up the playing field for the 21st Session in 2015 in Paris, discussed in Chapter 15, Section 15.6. The AR5 urges policymakers, as well as the public as a whole, to cooperate globally, stating evidence of climate turmoil, proposing future trends, and offering adaptations and mitigation options to combat the climate change issue.

Chapter Summary

As the history of climate change shows, earth has undergone numerous periods of warming and cooling in its past. Earth will survive and adapt to another period of warming or cooling by natural internal, or external, forces; however, the different species and living organisms on earth may not. Dramatic growth in civilizations and species is very dependent on atmospheric stability and consistent climate. Humans have become a very adaptable species, but whether we can endure an upcoming major climate change will depend on just how adaptable we can be (Cullen, 2010, p. 12). Over earth's lifetime, countless species have become extinct due to climate change, whether it be a cyclical cycle or due to an outside force. Studying these species, along with tree rings, ice cores, and various other tools and methods, allows researchers and scientists to best interpret what occurred years ago, and what is occurring today. Many techniques used to study climate change from the past up until today are not without inaccuracies. However, there is an ever-growing support of data showing that the climate is changing globally. Various reports and research, such as the AR5 and the Copenhagen Diagnosis, were written or conducted in order to get a better understanding of the state of the earth's climate, and where it is likely to be heading. Bringing the world together to discuss possible solutions and theories is a great step that had to take place in order to get a better and more accurate understanding of this planet's climate.

References

Book References

Ahrens, C.D. (2007) *Meteorology Today: An Introduction to Weather, Climate, and the Environment*, Eighth edition, Thomson Brooks/Cole, Thomson Higher Education, Belmont, CA.

Allison, I., Bindoff, N., Bindschadler, R., *et al.* (2009) *The Copenhagen Diagnosis*, UNSW Change Research Centre, Sydney, Australia.

Cullen, H. (2010) *The Weather of the Future*, HarperCollins Publishers, New York.

Rafferty, J.P. (2011) *Glaciers, Sea Ice, and Ice Formation*, Encyclopedia Britannica Inc., New York.

Journal/Report References

Eldredge, S. and Biek, B. (2010) Ice Ages: What Are They and What Causes Them? *Utah Geological Survey*, http://geology.utah.gov/map-pub/survey-notes/glad-you-asked/ice-ages-what-are-they-and-what-causes-them/. 10 July 2016.

"IPCC Factsheet: What is the IPCC?" (2013) *Intergovernmental Panel on Climate Change*. WMO. Geneva.

IPCC (2014) *Climate Change 2014: Synthesis Report*. Contribution of Working Groups I, II and III to the Fifth Assessment Report of the Intergovernmental Panel on Climate Change (Core Writing Team, R.K. Pachauri and L.A. Meyer (eds.)). IPCC, Geneva.

Riebeek, H. and Simmon, R. (2005) *Paleoclimatology: Written in the Earth*. Earth Observatory, NASA.

Stevens, T., Thomas, D.S.G., Armitage, S.J., *et al.* (2007) Reinterpreting climate proxy records from late Quaternary Chinese loess: A detailed OSL investigation. *Earth-Science Reviews*, 80, (1–2), pp. 111–136. Abstract.

Website References

"All About Frozen Ground." (2016) *National Snow & Ice Data Center*, https://nsidc.org/cryosphere/frozenground/methane.html. 11 July 2016.

Andrews, T. (2008) *Orbital Dynamics*, Climate Science: Investigating Climatic and Environmental Processes, NOAA, http://www.ncdc.noaa.gov/paleo/ctl/clisci100ka.html. 9 July 2016.

Bagley, M. (2013) *Holocene Epoch: The Age of Man*, Live Science. 27 March 2013, http://www.livescience.com/28219-holocene-epoch.html. 9 July 2016.

"Climate Basics." (2014) *Climate Prediction.net*, http://www.climateprediction.net/climate-science/basic-science/. 6 July 2016.

"Climate Change: How do we know?" (2016) *Global Climate Change: Vital Signs of the Planet*, NASA, http://climate.nasa.gov/evidence/. 21 July 2016.

"External Forcing Factors." (n.d.) *Climate, Adaptation, Mitigation, & E-Learning*, National Council for Science and the Environment, http://www.plantsciences.ucdavis.edu/plantsciences_Faculty/Bloom/CAMEL/external.html. 15 July 2016.

"How Do We Determine Past Climate?" (2016) *NIWA Taihoro Nukurangi*, https://www.niwa.co.nz/climate/faq/how-do-we-determine-past-climate. 9 July 2016.

Linacre, E. and Geerts, B. (1998) *Borehole Temperatures and Past Climates*, University of Wyoming, http://www-das.uwyo.edu/~geerts/cwx/notes/chap03/borehole.html. 11 July 2016.

Lovejoy, S. (2014) *Is Global Warming a Giant Natural Fluctuation?* Live Science, 18 April

2014. McGill University, http://www.
livescience.com/44950-global-warming-
natural-fluctuation.html. 13 July 2016.

Maasch, K.A. (1997) *What Triggers Ice Ages?* PBS,
NOVA, http://www.pbs.org/wgbh/nova/earth/
cause-ice-age.html. 10 July 2016.

Mason, M. (2016) *Dendrochronology: What Tree
Rings Tell Us about Past and Present*,
Environmental Science.org, http://www.
environmentalscience.org/dendrochronology-
tree-rings-tell-us. 10 August 2016.

"Milankovitch Cycles and Glaciation." (n.d.)
Indiana University Bloomington, http://www.
indiana.edu/~geol105/images/gaia_chapter_4/
milankovitch.htm. 9 July 2016.

"Milutin Milankovitch (1897–1958)." (n.d.)
Orbital Variations, NASA Earth Observatory,
http://earthobservatory.nasa.gov/Features/
Milankovitch/milankovitch_2.php. 9 July 2016.

"Organization." (n.d.) *Intergovernmental Panel on
Climate Change*, WMO & UNEP, https://www.
ipcc.ch/organization/organization.shtml.
12 July 2016.

Owen, J. (2010) *New Earth Epoch Has Begun,
Scientists Say*, National Geographic News,
6 April 2010, http://news.nationalgeographic.
com/news/2010/04/100406-new-earth-epoch-
geologic-age-anthropocene/. 9 July 2016.

Pewe, T.L. (2016) *Permafrost*, Encyclopedia
Britannica, https://www.britannica.com/
science/permafrost. 11 July 2016.

"Precession." (n.d.) *Global Climate Change*,
Section 2.5.2.3, http://www.global-climate-
change.org.uk/2-5-2-3.php. 10 July 2016.

"The IPCC: Who Are They and Why Do Their
Climate Reports Matter?" (n.d.) *Union of
Concerned Scientists*, http://www.ucsusa.org/
global_warming/science_and_impacts/science/
ipcc-backgrounder.html#.V5l7N6KM755.
12 July 2016.

"Topics in Studying Climate Change." (n.d.)
Global Change Project, Paleontological
Research Institution and its Museum of the
Earth, http://www.priweb.org/globalchange/
climatechange/studyingcc/scc_01.html.
9 July 2016.

"UN Climate Change Newsroom." (n.d.) *United
Nations: Framework Convention on Climate
Change*, http://newsroom.unfccc.int/about/.
12 July 2016.

"What Are Greenhouse Gases?" (n.d.) *Greenhouse
Gases*, NOAA, https://www.ncdc.noaa.gov/
monitoring-references/faq/greenhouse-gases.
php. 11 July 2016.

"World Data Centers." (n.d.) *NOAA & NCDC*,
https://www.ncdc.noaa.gov/customer-support/
world-data-centers. 10 July 2016.

Zimmermann, K.A. (2013) *Pleistocene Epoch:
Facts about the Last Ice Age*, Live Science.
9 October 2013, http://www.livescience.
com/40311-pleistocene-epoch.html. 9 July 2016.

**Figure References – In Order
of Appearance**

Lemke, K.A. (2016) *Causes of Climate Change.*
University of Wisconsin-Stevens Point-Geology
370, https://www4.uwsp.edu/geo/faculty/
lemke/geol370/lectures/12_climate_change.
html. 18 August 2016.

Buxbaum, T., Thoman, R., and Romanovsky, V.
(n.d.) Achieving the NOAA Arctic Action Plan:
The Missing Permafrost Element. Alaska
Center for Climate Assessment and Policy.
University of Alaska Fairbanks, p. 4.

12

Extreme Weather

Nicole Gallicchio

Everything in existence tries to be in homeostasis, creating a simplistic balance. Earth and its weather are no exceptions from this consistently working behavior. There are influences and elements on earth that cause the atmospheric balance to be altered. When the atmospheric balance is severely altered, extreme weather can result. There are numerous ways to define extreme weather, and many different classifications. Climate change will also influence extreme weather and its future trends. Current climate trends and constantly changing atmospheric conditions have led to recent changes in extreme weather.

12.1 Current Climate and Weather Trends

When discussing extreme weather, the first thing that has to be examined is the current state of the climate. While we cannot be 100% certain on the accuracy in past climate data, we can be as accurate as possible with current conditions. Current instruments – such as satellites, weather stations, climate monitoring stations, and various other atmospheric devices – allow for the success of current climate studies. These scientific advancements allow us to measure trends in all layers of the atmosphere and oceans. Extensive data can show us about the current state of our climate and the trends in today's weather patterns. Many are confused about the difference between weather forecasting and climate forecasting. Weather occurs today; climate occurs over a period of time. Weather forecasts are able to take atmospheric conditions, input them into computer models, and output projected data. Climate forecasting models are geared to produce trending data, such as increases in temperature and moisture. These models are unable to provide information such as "it is going to snow 10 inches on this specific date 20 years from now," due to the error and uncertainty in the changing atmospheric elements. In layman's terms, the weather forecast is never 100% accurate, due to the amount of unaccountable factors of atmospheric elements that technology cannot measure. This is often very difficult to relay to the public. The public witnesses the amazing achievements and advancements of mankind every single day, such as the ability to go into space, or even cure cancers. Humans and technology have the ability to do all these amazing things, yet we still cannot be 100% sure what tomorrow's weather will be.

When taking a look at earth's most recent climate, we can best pinpoint the changes in weather patterns due to the documentation that

The Evolution of Meteorology: A Look into the Past, Present, and Future of Weather Forecasting, First Edition.
Kevin Anthony Teague and Nicole Gallicchio.
© 2017 John Wiley & Sons Ltd. Published 2017 by John Wiley & Sons Ltd.

has accumulated over the last 5000 years. Examining the elements that are associated with weather, one can look at each individually. Temperature can be examined both regionally and globally. Temperature change in one specific region may cause a more drastic result than it does in another region, depending on the type of region, landscape, and ecosystem. Technology today is largely more precise than it was 20, 30, even 50 years ago (Figure 12.1). Current weather data are more accurate than they have even been, and will only increase in precision. A significant number of data show an increase in temperature over the last several decades, with a focus on the past 20 years. An increase in temperature naturally creates a change in weather trends. Global weather patterns have recently began to shift/fluctuate, despite the argument of their root cause. Different areas of the world have experienced different weather anomalies. Some regions have experienced more frequent drought, while others increased flooding. Some regional areas have seen a more temperate climate with unusual

spikes in weather phenomena. We can look at two specific types of weather trends: global weather trends and local/regional weather trends. These two weather trends create two different scenarios. Global weather changes over time and for the most part it happens gradually. Local weather changes and trends seem to happen more rapidly. This rapid change may just be because people's lives are closely affected by local changes and phenomena that occur. For example, if people in an area experience a large amount of flooding for five years in a row, they will feel as if their immediate climate is changing, and therefore may have the tendency to say climate change is occurring rapidly. Society has a tendency to base its beliefs and theories about climate change and weather extremes on local experiences. While society needs to consider small-scale weather trends in order to protect lives and take precautions, scientists need to analyze weather and the atmospheric trends on a global scale, in order to attempt to preserve earth's environment for sustained life to continue.

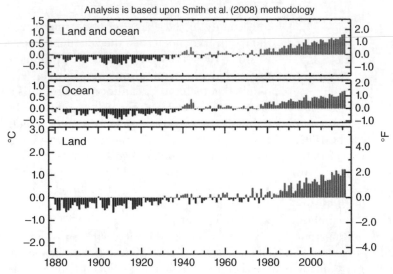

Figure 12.1 Temperature swing from the 1880s through the turn of the 21st century on land and ocean. There is clearly a warming trend based on these data over both the land and oceans. *Source*: Courtesy of NOAA Photo Library.

Earth's atmosphere plays a vital role in regulating the surface temperature by providing a layer of gases that protects us from excessive heat and harmful radiation from the sun. The atmosphere also traps heat rising from the earth's interior, providing further warmth. This leads to the current controversy surrounding the debate regarding the warming or cooling of earth's global average temperature and greenhouse gases. There are scientists who believe in global warming, there are some who believe we are in a little ice age, and there are some who feel we are in a natural cycle. Conflicting beliefs arise regarding solar activity and its relation to the temperature and fluctuations of earth's natural atmospheric cycles. There are debates surrounding current climatic patterns and global meteorological anomalies and whether they are caused by human activity. There are some scientists who believe if they go against saying that human activity is to blame for climate change that they will be cast out from the scientific community. However, science is all about continuous learning. Asking questions and testing theories is how science advances. No harm can be done by attempting to disprove a scientific theory: it can only lead to further proof of the theory, or corrections of mistakes that have been made, in turn furthering and opening more doors for the facts of the theory.

As discussed throughout the book, temperature influences all atmospheric processes. Today, temperature values are typically measured in anomalies, the prime example being of the hottest temperature ever recorded on earth in 1913, in Death Valley, in the United States, of 134.06 °F (56.7 °C). Scientists are always looking at today's temperatures and trends and asking whether they are abnormal for a specific time and/or location. All-time record temperatures have recently been recorded in several places around the world. A few recent record-breaking temperatures have recently occurred in areas in which high temperatures are part of the norm. India is a well-known region for high temperature values, with a variety of climates regions throughout the country. A large majority of its climate contains humid, tropical, and warm climates. On 19 May 2016, a heatwave swept through the desert state of Phalodi, India, leading to a high temperature of 123.8 °F (51 °C), breaking the record of 123.08 °F (50.6 °C) set in 1956. Thailand, Laos, Cambodia, and the Republic of the Maldives were also affected by this massive heatwave that crossed this region of northern India. Temperatures exceeded 104 °F (40 °C) for weeks. Although this area is characterized by weeks of sunshine and increasing heat, temperatures above 122 °F (50 °C) are abnormal ("India Records Its Hottest Day Ever," 2016).

Another area of the world that is accustomed to very high temperatures is the southwest region of the United States. More recently, however, on 19 June 2016, Tuscan, Arizona reached a scorching 105 °F (40.6 °C), while Phoenix, Arizona recorded a temperature of 118 °F (47.8 °C), and Southwest Uma, Arizona recorded 120 °F (48.9 °C). All of these cities came within the top five hottest ever recorded for the area during this heatwave. On a small scale these current weather changes and spikes in temperature are significant; however, on a large scale they are diluted. As previously stated, written records only make up a small portion of earth's history. It seems unreasonable to speculate that cyclical changes in earth's temperature would not be possible. More data collection and research is needed to provide a definite answer to whether recent temperature spikes are correlated to a specific function, or if they are a part of natural changes in earth's cycles. Earth is a living body that responds to its ups and downs, reacting accordingly to maintain balance and homeostasis.

12.1.1 Global Temperature Anomalies

Global temperatures have been at the forefront of scientific research for decades. Paleoclimate research has led to the tracking of climate temperatures since before the recording of weather. Even though these temperatures have a large margin of error and uncertainty, we can still gauge trends. Temperature anomalies are defined as a deviance from the long-term temperature average, or reference value. These anomalies can be used to describe land, air, or sea surface temperature values. The National Oceanic and Atmospheric Administration (NOAA) has output that June 2016 marks 14 consecutive months of record heat for the globe with the average sea surface temperature at a record high. The NOAA archives date back 137 years to 1880, and it concludes that the average global temperature for 2016 was 1.89°F above the twentieth-century average ("June Marks 14 Consecutive Months of Record Heat for the Globe," 2016). The average global sea surface temperature for the year to date was the highest for January to March 2016 than it was throughout the 137-year history, according to the National Centers for Environmental Information of the NOAA. It has been noted that record warmth has been observed in various areas around the globe, encompassing every continent (NOAA National Centers for Environmental Information, State of the Climate: Global Analysis for March 2016, 2016). The causes of these anomalies are constantly scrutinized and debated.

12.2 Classifying Extreme Weather

How do we classify one weather event as being more extreme than another? As a global society, we classify extreme weather on a small scale in time in relation to weather phenomena throughout earth's existence. The definition, classification, and diagnosis of extreme events are far from basic. There is no universal unique definition of what an extreme event is. Extreme weather is typically defined today as weather events or phenomena that are severe, unseasonal/rare, or out of the normal range. Severe weather is defined as meteorological phenomena that could potentially cause damage, social disruption, devastation, or loss of human life. The severity of these phenomena is typically measured on long-term loss, and human state of affairs. Unseasonal weather is defined as weather that is not typical for a specific area's current season. Rare events are events that are unlikely to occur. Weather that is out of the normal range is associated with elements that have data recorded above or below average amounts. Extreme and severe weather can be classified by each atmospheric element being different from its normal occurrences (Stephenson, 2008, p. 11).

As the climate changes, the processes that drive extreme weather will naturally change accordingly. We see this with today's climate and the different oscillations that occur, their predictability for fluctuations of extreme weather, and its dependency on the scale of the oscillation. The slightest change in any or all processes can spark an extreme weather event that earth has not seen before. As time goes on, no matter how the climate changes, new extreme weather patterns will arise. As a society, we can become aware of what conditions influence extreme and disruptive meteorological conditions, and how to identify the precautions that need to be taken in order to protect ourselves and our environment. As scientists, understanding each atmospheric mechanism and its potential connection to extreme weather phenomena can help with the predictions of future meteorological patterns of this type of weather occurrences.

12.2.1 Thresholds

Extreme weather events can be very multi-dimensional. Typically, when extreme events are labeled they are solely based on one attribute instead of many, such as spatial scale, time and duration, rate, intensity, and multivariate dependencies (Stephenson, 2008, p. 13). Often we can begin to group and classify extreme weather by the type of event. For example, such events may include tropical cyclones, convective and mesoscale phenomena, flooding and drought, temperature spikes and dips, and fog. The Intergovernmental Panel on Climate Change (IPCC) defines "complex extreme" events as "severe weather associated with particular climatic phenomena, often requiring a critical combination of variables" (Stephenson, 2008, p. 13). Typically, when thinking "extreme," common definitions include a maxima and a minima. Choosing a threshold in order to define the extreme events can be a proficient way to determine extrema. The simplest way to begin this approach is to choose an absolute constant threshold that relates to old impacts. This will allow a single value to be maintained to compare an astray from relative thresholds. A relative approach is to choose a constant relative to a distribution function of variables at specific locations. This approach ensures that a given fraction of events will by definition be "extreme." It can classify events as rare or "record-breaking" when choosing a threshold to be the maximum value, or number of occurrences of all previously observed values (Stephenson, 2008, p. 18). Extreme events can be classified as having exceedances above a high threshold. This type of classifying is susceptible to various types of statistical analysis. Exceeding occurrences at irregular times and the excess tend to be strongly skewed due to not having a controlled time series analysis (Stephenson, 2008, p. 16). However, exceedances can be considered a realization of a process called a "point process." The point process technique can be a useful tool in characterizing properties of simple extreme events (Stephenson, 2008, p. 22). Point process is a technique that uses random occurrences seen in data in order to isolate a grouping, typically seen in meteorology for notifying points in time or geographical space. An example of this process is the occurrence of lightning strikes when they are recorded according to their location in time and geological space. In relation to extreme weather events it can be used to characterize properties such as rate, magnitude, and temporal clustering.

12.2.2 Temporal Duration

The temporal duration of extreme events is an important part when classifying. It is extremely significant when classifying the difference between climate events and an extreme weather event. A climate event is dependent on duration or frequency of an event, typically in this respect the duration is a chronic, elongated, or frequent event. A chronic event has a long duration. An acute extreme event is one of severity but has a short duration. Acute extreme events typically are how weather systems are classified. An example of a chronic event that could impact climate would be an extreme drought, or a long heatwave. These types of events create long-term damage and impacts to society, the ecosystem, and overall atmospheric conditions. While there is no dismissing the severity and damage that can be done by an acute event, long-term climate events affect on a larger scale.

12.2.3 Climate Extremes Index

"Observed changes in extremes should be considered in light of observed changes in mean quantities, including observed changes in annual average temperature, and changes in maximum and minimum temperatures and the diurnal temperature range (DTR)" (Easterling, 2007, Abstract). The normal ranges or average weather amounts are currently taken from recorded history values. These values have only accounted for a small percentage of weather occurrences of earth's existence. Seeing as how recorded meteorological events only date back a couple hundred years, it is hard to accurately speculate the strengths and weaknesses of weather events. Today's current society defines weather to be "extreme" due to the repercussions it has on the ecosystem, human life, and society (Figure 12.2). Weather has the power to destroy, devastate, and damage on a large scale.

In order to better measure and relay the strengths of extreme weather events, the NOAA devised an index for climate extremes in the United States, the Climate Extremes Index (CEI). The method for determining climate indices was introduced in 1996 with the goal of summarizing and presenting a complex set of multivariable and multidimensional climate changes, in order for the results to be understood and used in policies by nonspecialists ("US Climate Extremes

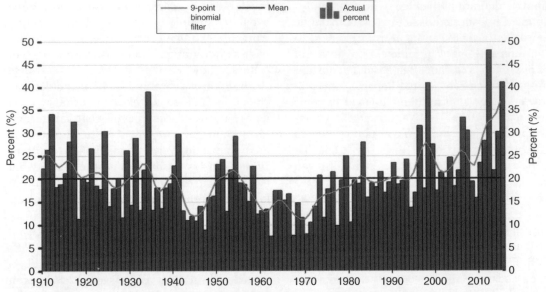

Figure 12.2 The NOAA's Climate Extremes Index (CEI) helps categorize extreme weather events. This is a valuable tool when trying to measure qualitatively and quantitatively the pattern of occurrences of such events. This figure shows a period of extreme events in the United States from 1910 through 2015. While there is some wavering throughout this period, there has been an increasing trend in extreme events since the 1970s, with a fairly rapid increase since the 21st century. *Source*: Courtesy of NOAA Photo Library.

Index (CEI): Introduction," n.d.). The CEI is the arithmetic average of five to six indicators of the percentage of the continental United States. It factors in such data as percentages of maximum and minimum temperatures below or above the normal range. The normal range can be defined as the conditions falling in the upper and lower tenth percentile of the local period of record. It calculates the percentage of severe drought or severe moisture surplus based upon the lowest or highest tenth percentile according to the Palmer Drought Severity Index (PDSI). The PDSI uses recent precipitation and temperature data to calculate water supply and demand. It incorporates soil moisture and primarily reflects long-term drought/dryness ("Drought.gov," 2016). When looking in a long-term variation, or change in the tendencies, it can provide insight into the increase, decrease, or steadiness of extremes of climate. This output can help

classify extremes in weather from determining the extremes in climate variations.

12.2.4 Using the Past for Today

Scientists can also use ancient writings and literature to correlate scientific predictions with translated accounts. Translated language can help interpret observational accounts of weather. Collectively with paleoclimate data, these translated documents can assist in the recreation and analysis of past weather patterns. These accounts and data can help us classify today's extreme weather by comparing it to past weather events. If scientists can illustrate strength in weather events by comparing data such as wind, precipitation, and pressure levels, there will be a foundation for benchmarking weather events. Fossils also are a tool for scientists, as they contain information about ocean temperature, chemistry, currents, and surface winds. Constructing data from different sources using microfossils as a source can provide information regarding wind and weather patterns. "The most valuable fossils found in sediment cores are from tiny animals with a calcium carbonate shell, called foraminifera" ("A Record from the Deep: Fossil Chemistry," n.d.). These different techniques aid in classifying extreme weather by providing further documentation and evidence into past weather conditions. Further additions to current data only create a more accurate classification.

12.3 Extreme Weather Influences

If we understand different processes that lead to the creation of extreme events, we can perhaps predict how they will change in the future, and improve our overall understanding, knowledge, and prevention capabilities. Extreme weather is influenced and fueled by several atmospheric and oceanic factors. These extreme weather events also have influences and impacts on human life and the environment. The extent of weather depends on variations in the atmospheric elements. The two major causes of weather variations are temperature and moisture. These two atmospheric conditions are also influenced by different outside conditions, such as the sun's energy, human activity, and the rotation and orbit of the earth.

12.3.1 Temperature

Temperature is a measure of energy at an atomic level, or more commonly the degree or intensity of heat in an object or substance. Typically, temperature is thought of in degrees based on a comparative scale; however, when broken down to its purest sense, it is how much energy and speed there are of atoms in a given area. These atomic molecules do not all move at the same speed, and it is partially dependent on the type of molecule and its physical state. Temperature correlates to moisture in

numerous ways. One way is the sense that the amount of moisture the atmosphere can hold directly relates to the temperature and movement of atoms. Temperature and moisture both create atmospheric instability; these unstable fluctuations cause an upset in atmospheric balance resulting in implemented weather conditions. These changes also impact the atmospheric flow, or jet stream, which is dependent not only on earth's orbit and rotation but also on hot and cold air boundaries. When you analyze each factor that contributes to extreme weather and breakdown their influences, you realize that they all rely on and are subject to the temperature and temperature changes. Temperature change influences extreme weather because it is a primary contributory factor to all atmospheric processes. When the atmosphere is warmed by temperature increases, it is able to hold a greater amount of water vapor. This increase in water vapor, in turn, increases the warming of the atmosphere. As temperature increases, water molecules gain energy and are quick to evaporate. More evaporation leads to more moisture in the atmosphere, which causes an increase in heat, which is radiated back to earth. This is due to the fact that as water evaporates it releases heat into the atmosphere. This increase in water vapor also leads to more precipitation. A furthering of precipitation only fuels extreme weather.

12.3.2 Convection

Convection is a form of heat transfer by the mass movement of a fluid. Fluids such as liquids and gases are susceptible to convection, due to their ability to move freely and potential to develop currents. Convection is a naturally occurring transfer in the atmosphere. This type of heat transfer within the atmosphere enables the development of rising and cooling air. As the earth's surface heats, the air molecules adjacent to the surface heat through conduction and gain more energy. The earth's surface heats unevenly, causing a temperature gradient in adjacent air molecules. The heated air molecules become less dense than the cooler air molecules, and the heated air molecules expand and rise. The air at the surface is then replaced by the air that arose and cooled, which then in turn is heated by the surface. This vertical interchange of air molecules is known as "convection" (Ahrens, 2007, p. 33). When localized convection is coupled with atmospheric instability, storms are highly likely to develop. Such conditions that influence atmospheric instability are parcels of air that are warmer than the surrounding air, making them less dense, which lifts the parcels further. It should be noted that dry and moist atmospheric conditions affect instability differently. Depending on the saturation of the air parcel, the likelihood of instability can be calculated through environmental lapse rates. As the air temperature drops rapidly with increasing height, instability increases. These conditions can arise through several factors, not limited to but including cold advection (winds bringing in colder air), radiational cooling, and daytime solar heating. These instabilities that are created

directly create storms and turbulence in the atmosphere. Once these atmospheric disruptions are coupled with moisture they can lead to severe weather events.

12.3.3 The Hydrologic Cycle

Salinity, temperature, and weather are all an interconnected phenomena. These are all connected through the water cycle (or hydrologic cycle), the circulation water takes as it moves from the earth's surface into the atmosphere and back again (Figure 12.3). Ocean salinity is important because it affects ocean circulation, which in turn affects ocean temperatures, which then can alter weather. High salinity will cause the salty water to sink and the warmer water to rise to the surface, impacting ocean currents. Changes in ocean currents and warmer waters affect weather patterns and strength of weather events. The warmer the water is, the more energy and added fuel will be available for a

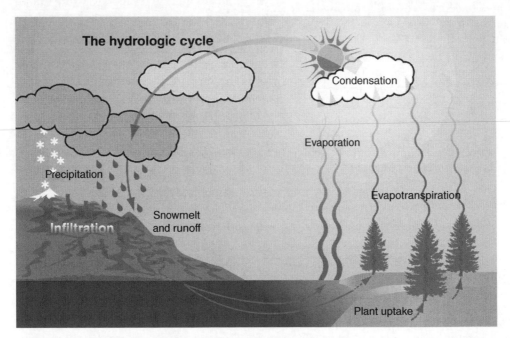

Figure 12.3 The hydrologic cycle is a cycle of events concerning the movement of water from the atmosphere to the surface of the earth and back to the atmosphere. Changes in climate may have great impacts on this cycle. If any part of the cycle is greatly disrupted, the cycle will be changed or even stopped in some locations, resulting in devastating and life-changing climatic conditions. *Source*: Courtesy of Delaware River Basin Commission (www.drbc.net).

storm to develop and strengthen. The warming of the ocean increases evaporation and therefore pulls more water vapor into the atmosphere. This phenomenon is one that is accrued over time due to slow changes that occur to warm the ocean, as well as the addition/removal of fresh water, and both aspects alter salinity levels. This, however, is not the only way that the oceans can influence severe weather.

12.3.4 Ocean and Tidal Cycles

Danger to human life can be exacerbated when high tides coincide with severe weather. High lunar tides can significantly increase flooding in coastal areas when severe weather strikes a region. Storm surges and rising water levels due to large amounts of precipitation and strong winds create damaging effects on the ecosystem, land formation, and human life. There has been research looking into the correlation between intensity in weather patterns and lunar tides. The research presents theories on the pressure variations caused by tidal oscillations. Lunar tides are at their strongest when the ocean's surface is closest to the moon or opposite, and this makes the ocean bulge outward. Ocean tides are at their weakest when the moon is perpendicular to any specific location, causing low tides. Part of this theory is contradicted by the complications that are associated with oceans, such as uneven water depths and land presence. However, the atmosphere can be theorized as a large body of water that does not associate with uneven depths or surface interruptions. Newton's theory of gravity provided correct information correlating the phases of the moon and was used a century later to develop our understanding of atmospheric tides. Atmospheric tides are generally described as planetary-scale oscillations driven by solar heating periods, influenced by integral fractions of a solar or lunar day. These tides are detectable by variations in air pressure and should be further explored on a global modeling scale.

Now that we have defined oceanic and atmospheric tides, the question is whether they can affect the strength of meteorological elements. Ocean tides can strongly enhance damage caused by extreme or severe weather. When an astronomical tide coincides with a strong storm, sea water levels will be higher, creating a greater threat of damage and devastation to ecosystems and society. Atmospheric tides had been under observation for several decades for their potential effect on weather. In January 2016, the Geophysical Research Letters published a paper that proposes atmospheric tides could alter precipitation rates along with the increase in atmospheric pressure (Kohyama and Wallace, 2016). Data revealed that during atmospheric high tides rising air pressure slightly increases air temperature. In turn the air temperature boost allows the air to be able to hold more water vapor, lowering the humidity levels, leading to a decrease in the probability of rain. Conversely, during a low tide, pressures drop slightly, lowering the air temperature, raising the humidity, and increasing the probability of rain (Sumner, 2016). Whether we are discussing atmospheric or oceanic tidal events, it will not directly affect the magnitude of a weather event in respect to direct atmospheric elements and processes. Tides, however, can be a contributing factor to the devastation that may occur.

12.3.5 Ocean Current Change

Ocean currents act as a global conveyer belt, moving water around the globe. They transport enormous amounts of heat around the world, making them one of the most important driving forces of climate. Ocean currents can be generated by wind, density differences in water masses caused by temperature and salinity variations, gravity, and events such as earthquakes ("Ocean Explorer," 2013). Winds blowing over the surface of the sea will produce ocean currents. Ocean circulation comprises a global network of interconnected currents, counter-currents, deep water circulation, and turbulent eddies (Schiele, n.d.). Water plays a central role in earth's climate. Its composition also plays an important role in the climatic system. The density, salinity, and temperature of earth's water alter how the atmosphere interacts with the ocean.

Water vapor provides the strongest warming effect of all the greenhouse gases. Water, as a fluid, covers nearly 70% of the earth's surface. The oceans in large part respond to winds, which make the surface susceptible to climate change and further influencing change both atmospherically and meteorologically ("Ocean Circulation & Climate," 2002). Any change in ocean currents has a large effect on earth's atmospheric and meteorological conditions. The ocean absorbs and stores more heat than the atmosphere. Both the atmosphere and ocean transport heat, although the ocean transports heat slower than the atmosphere. Multiple forces keep thermohaline circulation in perpetual motion below the surface. Thermohaline circulation is the process of deep ocean currents that are driven by changes in water density, which is dictated by salinity and temperature. Surface height differences affect atmospheric circulation. If you change the chemical composition of any mixture, compound, or element, it will not behave the same as it did before. Therefore, if you change the chemical composition of earth's oceans, its corresponding balancing current cycle will be altered. PH levels, acidification, salinity levels, and the effects of changing marine life will all have repercussions on each other if any of them are altered, causing a chain reaction in order for the ocean to try to rebalance itself.

There are theories regarding the amount of freshwater running off from ice caps entering saltwater oceans. Ideas surround the mixing of the two and the effects it could have on ocean currents and therefore atmospheric circulation. Two examples lie within the possible changing of ocean processes. One example is the changing of ocean current strength and direction. The second is the balancing rate of evaporation, due to salinity levels decreasing with the addition of fresh water and an increase in evaporation from the overall increase of the earth's temperature. The importance these theories could potentially have for life on earth remains a great unknown, prompting a great deal of debate.

Nearly 40% of earth's population lives near coastal regions, signifying that a large majority of the world's population is at risk for the most damage during an extreme weather event (Figure 12.4). This population is susceptible to rising water levels, changes in ocean currents, storm surges, erosion, flooding, tsunamis, and storms. A change in ocean currents has the potential to change the atmospheric patterns that society has become accustomed to, and shifting them to other regions. This shift would bring unusual weather to an area that might not be prepared for it, and when located near the coast it is apparent that it is easy for these areas to be devastated by such a change.

12.4 Is Extreme Weather on the Rise?

The written documentation of weather data dates back to ancient Mesopotamia in 3000 BC (5000 years ago), which is a small scale when compared to the age of earth. Therefore, documented weather accounts for only 0.0001% of earth's life. When looking on a large scale of earth's life, there have been at least five recorded ice ages since earth was formed 4.6 billion years ago. However, when asking the question "Is extreme weather on the rise?" several connections and correlations need to be made. The correlation between the natural changes in climate and the human influence on climate change is an important aspect in determining the cause and effect of extreme weather. Fluctuations in earth's climate, both past and present, need to be taken into consideration to study extreme weather and its relation to human influences and the cyclical repercussion of earth's ever-changing climate and atmosphere.

When an extreme weather event occurs, scientists can then analyze the event and recreate it step by step. This recreation allows scientists to examine the cause of the weather and evaluate the odds of the occurrence. These odds can tell us the likelihood of this extreme event

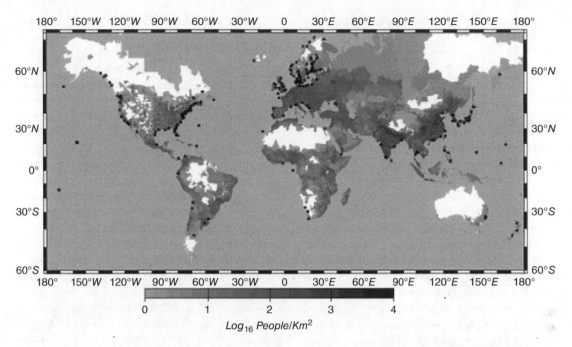

Figure 12.4 The global population based on the year 2000, with darkest shading representing the more densely populated regions. The darker shading is more prevalent along the coasts, showing that the world's population gravitates toward coastal locations. Also represented on this map is long-term sea level monitoring stations, represented by the black dots. *Source*: Reproduced with permission of American Association of Petroleum Geologists (AAPG).

striking again, along with the atmospheric conditions that need to be present for the event to formulate; in other words, a recipe for disaster. Recent statistics portray the rise of extreme weather over the course of the last couple of decades, or even years. Current statistics will also show tendencies toward more frequent and intense extreme weather in the future. These statistics are based upon model projections of global temperatures and increased human emissions. Since the most prominent models continuously factor in human emissions, it plays into the disbelief factor for some scientists and society.

Extreme weather by definition takes into account not only weather processes but also damage, loss of life, economic losses, and environmental destruction. As society expands, weather event devastation is only likely to increase. Put simply, the more people there are and the more money/property they have, the greater the possible damage. Analysis of weather processes alone is the only way to judge whether extreme weather is on the rise. All the other factors that are accounted for in extreme weather are consistently changing throughout earth's lifetime. The only control and constant variable is weather processes and elements.

If one were to research whether extreme weather was on the rise and compared data, one would find a significant amount of data supporting the theory that it was on the rise. Some look at loss of life as a characteristic of extreme weather. Appendix Figure C and D both show Loss Events Worldwide from 1980 to 2015 in relation to peril, with Appendix Figure E showing the 10 deadliest events in regards to Significant Loss Events Worldwide from 1980 to 2015. Others look at monetary figures and consider extreme weather is on the rise due to the increase in cost and damage that occurs during weather disasters. Appendix Figure F shows the

10 costliest events in regards to Significant Loss Events Worldwide from 1980 to 2015. It is necessary to dig deeper and look at trends in temperature, precipitation, and atmospheric pressure. An example of an increase in weather processes over the last decade can be seen in the United States, with the number of heavy precipitation events. According to the National Climate Assessment, observed trends in heavy precipitation (heavy precipitation events defined as a two-day precipitation total that has exceeded the average only once in a five-year period) have increased by nearly 40% since 1900 ("Extreme Weather," 2014). The driving mechanism behind this change is understood to be warmer air being able to contain more water vapor than cooler air. Extra moisture in the atmosphere results in heavier rainfalls. Where this example plays a major role is with tropical cyclone development and intensities. In order to correlate heavier rainfalls and moisture content directly with development and intensity of tropical cyclones, one would have to graph the amount of tropical cyclones by year, since the 1800s, that were over a category 3 on the Saffir–Simpson hurricane wind scale, or equivalent on other scales monitoring tropical cyclones, with correlating precipitation rates and magnitude. These scales all take into account wind speed. There is some criticism that these scales do not take into consideration other important factors, such as storm surges and precipitation rates. However, the stronger the winds, the lower the pressure system, and the more intense the storm will be, including storm surge and precipitation levels.

12.4.1 Is Climate Change Influencing Extreme Weather?

Over more recent years, on a global scale, record-breaking weather conditions have occurred; storms have been shown to be stronger and temperature differences greater. Appendix Figure G shows the geographical view of the Natural Loss Events Worldwide in 2015. For weather conditions to change in this type of

magnitude there needs to be underlying factors contributing to the changes. A changing climate will always lead to a change in weather conditions. How can we be sure that climate change is influencing more extreme weather? The answer is simple; because it is already happening. Earth's changing climate has become undeniable, whether or not there is a unified reason, influence, or explanation for its change. Slow and steady changes have been occurring over several centuries. The data collected over the past century indicated that weather is getting more extreme. There are far more droughts, stronger tropical cyclones, greater extremes in temperature, etc., and these greater threats are increasing in strength and frequency. It is, however, more difficult for the eye to pinpoint the fingerprint of climate change by looking at weather, than it is to see the fingerprint of climate change within the climate system (Cullen, 2010, p. 51). There are many ideas surrounding the current and future state of the climate. By dissecting current atmospheric conditions and extreme weather events, one may be able to pinpoint which current climate change condition is the main culprit in the increase in extreme weather events.

With extreme weather gaining more notoriety with the help of greater public accessibility, such as through different forms of communication and social media, things that were once never known or reported are now available to the public with ease and abundance. This notoriety plays a large factor in the overall feeling that extreme weather is rising now that the public can see first- or second-hand accounts of devastation caused by severe and extreme weather. The world can attest to a changing atmosphere and weather patterns that lead to a change in the severity of the weather. The split decisions lay within the reasoning behind the recent climatological changes. Now that we have declared that extreme weather is on the rise due to climate change, it is only logical to ask ourselves what we can do in order to better prepare for and forecast it.

Speaking to society on the issues of climate change and its impacts and repercussions needs to be done in a delicate manner. It is important to relay climate change information to the public in a clear fashion. Data and scientific explanations can be confusing and misleading to some parts of society, especially with newer and developing theories and beliefs. There are a great many disbelievers, skeptics, and naysayers to the ideas of climate change, human influence, extreme weather changes, and the science of meteorology. Scientists have many challenges when explaining not only the causes of climate change but also its impact on and risk to society and the ecosystem. Humans have adapted to climate change (be it warming up or cooling down) throughout history. In order for society to protect itself from climate change now, explanations of what type of conditions will be associated with a changing atmosphere are needed. Explaining what will happen and what precautions need to be taken will prevent loss of life and can perhaps preserve the species from extinction. In the event that society adheres to all the precautions and warnings that scientists provide against a changing climate, will it suffice to protect the human race from extinction?

12.5 Weather Pattern Predictions

It is assumed that our technology should be advanced enough to provide an accurate weather forecast that extends over 10 days. Weather occurs at every moment of every day, and affects our daily lives without our really even taking a second thought about it. We grab the umbrella or decide to leave the coat at home. For something that is constantly occurring, it is natural for the public to feel it should be predicted more accurately and over longer periods. It is hard to see how something as seemingly simple as rain can be of the utmost complexity for the meteorologist to forecast with precise accuracy. Many feel that if we can reach the distant moons and planets of outer space we should be able to have better modeling of atmospheric trends and weather patterns for here on earth. The topic of using past weather data and

cyclical patterns arises with the public, with the feelings that recorded weather patterns should be available for determining current and future weather. The question then arises as to why or how we can better predict temperature fluctuations, but not future weather forecasts. This is answered by explaining that temperature depends on thermodynamics and the energy balance of our climate system. Weather can be described and broken down into several elements. Some of these weather or atmospheric elements include temperature, precipitation, wind, pressure, and humidity. These elements combine to make weather extremely variable and unstable. Each atmospheric element has its own set of physical and chemical atoms and specific characteristics, all of which contain variables.

Technology at the present time cannot allow for an accurate weather forecast for more than 10 days out. Therefore, when discussing weather for the future, it is an overall weather trend, instead of numbers, data, and expected day-to-day observations. Weather predictions of the future come from factoring in what happens when such elements as temperature and moisture change within our atmosphere. If global temperatures continue to rise at the rate in which they are increasing now, humans can expect subtle changes in weather patterns, which in turn have great consequences.

Climatic trends are always changing and always up for debate between scientists. We can debate and theorize from known scientific processes, research, historical data, and meteorological behavior the weather patterns for different outcomes of future climates. All atmospheric processes are driven by temperature and have a direct effect on weather patterns; therefore, an increase or decrease in the earth's temperature will alter weather globally. Everything in existence tries to be in homeostasis, creating a balance. One of the most basic examples on earth can be seen by the cause of wind on a mesoscale. Air pressure is constantly moving from high to low in order to balance earth's atmospheric surface pressure. Temperature as well is always moving from cold to hot, in order to have a balance in density. On a synoptic scale, earth's Hadley cells and trade winds move cyclically within themselves in order to create a balance on a global scale. The same can be said about the ocean, with its ocean currents transporting warm and cold water in order to create a global balance. When the climate is changing, earth and its atmosphere has to work in different ways in order to create this homeostasis, sometimes leading to newer and unprecedented weather events and phenomena.

12.5.1 Possible Future Weather Patterns for a Warming Climate

A warming of temperatures, by as little as a few degrees, will have a significant impact on earth's weather patterns. Increased temperature will increase the frequency of hot days and decrease the frequency of cold days. Land temperatures will see the greatest change and increase in temperature, due to land having a lower specific heat than water. Specific heat is the amount of heat energy required per unit mass to raise the temperature of a substance by one degree Celsius. Water has a very high specific heat, which means that it will need to absorb a lot of energy in order to change its temperature. With land having a lower specific heat, it is susceptible to longer heatwaves, and increased drought. A rise in temperature will increase the rate at which ice melts at the Polar Regions, causing a rise in sea levels. Warmer atmospheric temperatures allow for more moisture to evaporate, leading to an atmosphere that can hold more water vapor. Presumably, larger atmospheric moisture content leads to larger weather systems, and stronger storms. The connection between an increase in temperature and an increase in the number of extreme storms is still being researched. However, we can speculate on the systematics behind an extreme weather event. A tropical cyclone is fueled by warmer sea surface temperatures. With an increase in atmospheric moisture, more rainfall and overall energy for the storm can be expected. This creates the potential for stronger, more intense storms. Tropical cyclone winds are driven by changes in pressure. The lower the pressure of a tropical cyclone, the greater the wind speeds. Low pressure forms when atmospheric circulation removes small amounts of atmosphere from a region. This typically happens along a boundary between cold and warm air, when airflow tries to reduce temperature variation. Uncertainty lies in whether a warming climate will increase global temperature variations. With a warming global climate, jet streams will waver, creating potentially stronger temperature gradients due to amplified troughs and ridges in the upper atmosphere. For example, if a jet stream forms a significant deep trough, artic air will flow downward toward the equator. Cold artic air will contrast with an only increasing equatorial temperature, creating a significant temperature gradient. A large temperature gradient significantly increases wind, pressure anomalies, and precipitation probability. Another example that is likely to be seen during a warming phase is the increased moisture and increasing thermal land–sea gradient. This will generally drive a rise in monsoonal rainfall on a seasonal scale. In this example, the energy from the India Ocean and various atmospheric and oceanic circulations drive rainfall pattern changes.

Another factor to consider with a warming planet is the salinity of the oceans. As temperatures rise, ice caps melt. The melting of fresh water from the ice may decrease the salinity of the surrounding oceans. At high latitudes, this melting ice cools the regional waters and sinks, in turn releasing its heat into the atmosphere and making for moderate winter climates in places like England (Schirber, 2005). Lower ocean currents are driven by density levels of the water, based on temperature and salinity levels. If waters warm due to climate change and salinity levels decrease due to the influx of fresh ice water, then one can only expect the lower ocean currents to alter in speed, intensity, and even direction. Currents circulate this fresh melted ice water, leading to potentially altered seasonal and long-term climates. These alterations can affect atmospheric processes and in turn influence such things as tropical cyclone formations, duration of droughts, strength of heatwaves, and overall weather patterns across the globe. Seasonal high-pressure systems, such as the Bermuda High over the Atlantic, could be weaker or stronger, more north or south, resulting in drastic changes synoptically across the world. Jet streams will be of differing intensities with different locations for troughing and ridging to occur over larger scales. Possible evidence of this type of thinking could be with the North American winters of the past two years, with extremely deep and long lasting troughs in the northeast and extreme ridges in the west, leading to cold and snowy conditions in the northeast and both record-breaking droughts in California and record-breaking warmth in Alaska and northwestern Canada.

Changing of the ocean currents will also alter the locations and strengths of high- and low-pressures stages across the world. A warmer climate will create a more wavering jet stream. As arctic air pulls down due to a trough lower than it would be now, this temperature gradient would be even greater. The amplitude of the streams will intensify, in ridges and troughs, which creates stronger vorticity and atmospheric spin, placing temperatures in places they normally would not be. An example is if the jet stream is altered semi-permanently seasonal temperatures will tremendously fluctuate with impacts on all aspects of life.

Another possible outcome of a warming climate could be in regards to weather systems developing over ocean waters. This type of scenario in a warming climate has already shown that storms will likely be of greater intensity. As an intense storm moves across the waters, cooler ocean waters and a more chaotic and moisture-starved atmosphere is left in its wake. With cooler waters, less moisture, and a chaotic atmosphere, storms following the original will be weaker or non-existent. This type of setup could lead to fewer overall storms, but with those that do form being very strong. While this idea may be less supported by a majority of scientists, it is a possible result of a warming climate. This idea, along with the infinite other questions and theories on what would be expected in a warming climate across the globe, shows that there is enough proof that the weather patterns that we have seen for the past few centuries will likely be very different, but how different and in what ways is still mainly a mystery.

12.5.2 Possible Future Weather Patterns for a Cooling Climate

Since humans have lived through an ice age before, we can estimate the type of weather trends that would be expected should a cooling climatic trend begin now. It is difficult, however, to say what would occur should a sharp cooling trend develop, or if there were current atmospheric factors that would impact a cooling climate in a different aspect than at previous times. We can only postulate ideas and hypotheses through our knowledge of the elements and weather models. It makes sense to begin with the knowledge of weather processes in their most basic forms.

The simplest thought about a cooling climate is that the overall atmosphere will be colder. This more scientifically correlates with global temperatures cooling and ice caps at the poles beginning to increase, along with the formation of glaciers. Permafrost will also begin to strengthen, thus solidifying atmospheric gases into the ground that would normally be exposed and driving an increase in temperatures. A colder climate will likely increase snowfall totals and depth globally, which will affect agriculture and human practices. This would have repercussions on food intake, plant life, and animal species. However, increased snowfall totals does not directly mean that precipitation and moisture levels will increase. Lower moisture levels and less atmospheric water vapor are possible outcomes in a colder climate. This is due to slow-moving molecules not being able to cross into the gas phase. Colder atmospheric temperatures also lead to cooler water/ocean temperatures. Cooler ocean and atmospheric temperatures can potentially lead to less cyclonic activity.

Tropical cyclone development relies on the transfer of heat from the ocean to the atmosphere through evaporation. If evaporation rates are lowered by cooler temperatures, this will hinder the beginning spark of a tropical cyclone. Less atmospheric moisture could lead to an overall decrease in global precipitation rates, leading to increased droughts, creating larger desert and tundra regions. These changes to the atmospheric elements, weather, and environment would significantly affect not only human life but the biosphere as well. A cooling trending climate will have just as many repercussions as a warming climate. Humans will have to adjust and plan accordingly, no matter what change occurs.

Perhaps one day, technology will be advanced enough to calculate and analyze all atmospheric elements, individual atoms, and molecules, and create an output of the weather scenarios that are likely to occur in a specific situation. Determining cyclical changes in global temperature changes accurately, and in a timely fashion, will perhaps one day be possible, enhancing the odds of human, animal, and plant life thriving in a changing climate.

12.5.3 The Next Ice Age

Earth goes through periods of ice ages, some proving to be stronger and more intense than others. These ice ages are typically cyclical. Therefore, it is relatively simple to predict when the next bout will come around. As discussed in Chapter 11, we already know what an ice age is and the debates on the current state of earth in respect to a current ice age. The last major glacial period occurred 11,000 years ago, during the Pleistocene epoch. "As only 11,000 years has passed since the last Ice Age, scientists cannot be certain that we are indeed living in a post-glacial Holocene epoch instead of an interglacial period of the Pleistocene and thus due for another ice age in the geologic future" (Rosenberg, n.d.).

There are also theories and beliefs that the changes in global temperature that earth is currently experiencing are a repercussion of an impending ice age. If the next ice age were to occur after a relatively significant global temperature increase then the amount of ice on earth's surface would increase during a glacial period. As temperature increases, the amount of moisture in the atmosphere is extremely likely to increase. This increase will bring about greater snowfall potential. More snowfall could allow for more accumulation of ice in the Polar Regions, which in turn would lower ocean levels. All these changes would alter the global climate and unanticipated changes would have the potential to arise.

During the last glacial maximum nearly 26 million square kilometers (16 million square miles) of earth were covered by ice. During this period, ice in the northern hemisphere covered earth as far south as Germany, Poland, Canada, and parts of the Middle East and the United States. In the southern hemisphere ice covered parts of Chile, Africa, and Southeast Asia. Weather during this period consisted of cold temperatures and increased precipitation mainly in the form of snow. New weather patterns and air masses formed, changing climate around the globe. It is estimated that ocean levels shrank by approximately 120 meters (394 ft) during the peak of the last ice age (Gornitz, 2007). Warm areas became cooler and drier than during an interglacial period. Rainforest areas no longer saw massive amounts of precipitation and warm temperatures. Most deserts expanded and became even drier than before, with a few exceptions, primarily due to shifts in airflow patterns.

It is difficult to know whether earth's next ice age will be similar to the previous one. Making predictions based upon the ice age that occurred over 11,000 years ago would require a significant amount of data, from before, during, and after, which included temperature, moisture content, atmospheric patterns, global wind patterns, ocean levels, and solar activity. Without these data we could only base our predictions on paleoclimatic data, current data, and our current knowledge of atmospheric processes.

We could use models to assist in predictions, however, technology factoring in all the past and present data and output a model resembling an ice age would contain significant error in its calculations due to the high uncertainty in its input and equations.

Our short history on earth and even shorter records of climate and weather prevent us from completely understanding the implications of climate change, and an increase or decrease in global temperature. It is certain, however, that any fluctuation, either an increase or a decrease, would have major consequences and alterations for all life on this planet.

12.5.4 Jet Stream Positioning

Weather is moved around earth through the movement of air. Circulation of the atmosphere is caused by pressure and/or temperature differences between two places. This movement of air on a small scale is experienced as wind. Atmospheric circulation on a global scale is described as a means of circulating thermal heat energy from the tropical region to the polar latitudes. This thermal heat energy is a consequence of the sun and is governed by the laws of thermodynamics. Three tropospheric cells on a global scale exist in both the northern and the southern hemispheres: the Hadley, Ferrel, and Polar cells. These cells balance atmospheric forces on the earth's surface. It is not only this oscillation that enables heat transport but also the interaction between the atmosphere and the ocean plays a large role. The Coriolis effect also results in the influence of atmospheric circulation. The Coriolis effect is the apparent path of an object moving longitudinally on earth, or the apparent deflections of a path of an object in a rotating coordinate system. On earth an object in the southern hemisphere will undergo a deflection to the left, and in the northern hemisphere the deflection will be to the right. This phenomenon occurs due to earth's rotation. Should the earth stop rotating, the atmosphere would circulate simply back and forth between the equator and the poles ("Convection, Circulation, and Deflection of Air," 2016).

As already mentioned, there are three tropospheric cell circulations around the regions of 30°N/S and 50–60°N/S in which temperature gradients are the largest. Wind increases as the temperature gradient increases, allowing these latitudes to therefore have strong winds in the upper atmosphere. Two classifications of jet streams exist: the subtropical and the polar jet. The polar jets are located around 50–60°N/S, while the subtropical jet streams are located around 30°N/S. Between the two jets lie the mid-latitudes. To the north is the polar jet, the boundary of arctic air, and to the south is the subtropical jet, separating the mid-latitudes from the tropical air mass and environment. The polar jet should be looked at more carefully for numerous reasons. The polar jet is a region of maximum upper-tropospheric flow, with upper air convergence and divergence. A warming Arctic decreases the temperature difference between the poles and the equator, as long as the temperature at the equator increases at a slower rate than the Artic. This smaller temperature gradient means weaker winds, consequently weakening the jet streams. A weak jet stream will likely meander in its path, as opposed to when the jet stream is stronger, resulting in it flowing in a straighter and more direct path. A weakened meandering jet stream is slower, with big troughs and ridges, causing a tendency for weather systems to linger. A stronger jet stream usually results in weather heading into a region from a more direct influence, such as is often the case in the United Kingdom with weather driven straight from the Atlantic due to a fast-flowing jet stream (Fry, *et al.*, 2010, pp. 40–41). The southern hemisphere's jet streams closely resemble the northern hemisphere's jet streams; however, the southern hemisphere's jet streams do not have as much variability and change in their troughing or ridging. This is because the southern hemisphere has much less landmass to influence such changes than does the northern hemisphere.

If these jet streams begin to weaken due to a deviating temperature gradient, weather patterns around the globe will slowly being to change. Droughts or monsoons could last longer, snowfalls and storms would be slow moving and more devastating, and air masses could affect the temperature of an area for longer. This leads to even more unpredictable meteorological patterns which will likely result in various alterations to the environment, landscapes, and the overall wellbeing of various species and plant life. Jet stream positioning could be destructive to many parts of the world, should it be altered. Changing the global upper air rotation will have repercussions and could potentially kick-start a chain of unusual weather occurrences.

12.5.5 El Niño and La Niña

El Niño–Southern Oscillation (ENSO) is an ocean–atmosphere interaction in which the El Niño event and the Southern Oscillation occur almost simultaneously. Even though these two events occur almost simultaneously, they have their own function and are completely independent of each other. El Niño and La Niña occur when changing wind patterns displace warm and cool water in the equatorial Pacific. Both El Niño and La Niña have global impacts. The replacement of cold and warm water leads to atmospheric temperature swings and changes in humidity. Usually, wind blows from east to west along the equator in the Pacific Ocean. This strong wind flow piles up water in the western part of the Pacific, while in the eastern part deep cold water is upwelled to the surface to replace this imbalance of water displacement. El Niño is an extremely warm episode in the eastern Pacific, which occurs in irregular periods of several years within a large area of the tropical Pacific Ocean. During an El Niño, the winds pushing the water become weaker, and a warm current of tropical water flows southward and replaces this cold water at the surface. The warming of the sea surface temperature also in turn affects the winds, weakening them even further.

El Niño and La Niña events transfer huge amounts of energy around the planet's surface. This energy distribution has an effect on global weather patterns. Altering weather patterns, steering storms and rainfall to new locations, these shifts in precipitation affect human life and the environment. El Niño brings, among other things, wet and cooler temperatures to the southern United States, warmer temperatures across India and Southeast Asia, and dry conditions to Indonesia. Although the temperature change occurs across the equatorial Pacific Ocean region, it has worldwide weather effects. La Niña, when equatorial Pacific Ocean temperatures are warmer over the western pacific, has in most cases an opposite global impact. It brings wet weather to Indonesia and a drier, warmer environment to the southern United States. El Niño and La Niña both have the potential to affect seasonal variations in temperature and moisture, affecting weather elements, trends, and conditions (Figure 12.5).

As the climate changes, all the factors that influence and drive El Niño and La Niña will alter. These different driving processes will change the way El Niño and La Niña currently operate. This will heavily rely on the way the ocean and atmospheric circulations are affected by climate change. The more significant the change is with global circulations, the more severe the change will be in El Niño and La Niña episodes. Temperature gradients will also play a large role in the alterations of these events. A swing in temperature is the primary driving force in all global circulations, continuing the cycle of this shift during this event. A warming climate will only enhance the effects of both El Niño and La Niña. A cooling climate will not necessarily increase or decrease the effects of El Niño or La Niña. It will, however, have an effect on the placement of its resulting meteorological characteristics.

In recent years, many like to directly connect El Niño with causing severe storms. It is unreasonable to attribute a single storm to one climatic influence, especially when severe storms are complex and variable; however, El Niño

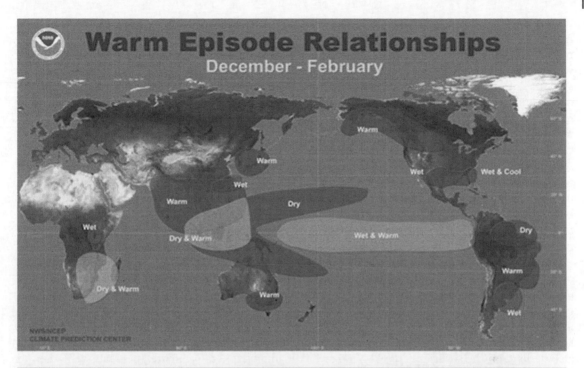

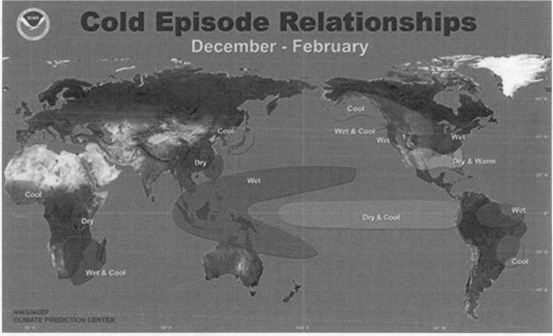

Figure 12.5 El Niño (upper image) and La Niña (lower image) events have drastic effects on the weather across the world, even though the root cause of these events stems from changes in ocean temperatures across the equatorial Pacific Ocean. *Source*: Images courtesy of the NWS/NCEP Climate Prediction Center.

fingerprints are possible to be present in some factors of a storm, dependent on the global region. El Niño typically creates conditions in which storms in certain areas are likely to form, and strengthen. During the El Niño or La Niña occurrence, the shift in precipitation can lead to major regional climate changes. When precipitation is shifted to new locations around the globe, it can bring severe drought, flooding, or storms to areas that are not accustomed to these weather events, in turn devastating society and uprooting the natural environment. Loss of life in this situation can occur for various reasons. Severe drought or flooding can lead to agricultural devastation, with the potential to result in extreme famine. Environmental changes can occur when precipitation is shifted from the norm. Wildlife and fish habitats can suffer loss or destruction, there can be increases of diseases in both animals and humans, and changes in landscapes can arise. The impacts on society also depend on the severity of the El Niño or La Niña occurrence. The stronger and more prominent the event is, the greater risk it imposes on society and the environment. Taking precautions is advised for these types of situations, to minimize the impact and devastation to habitats. Atmospheric and oceanic trends and predictions are essential to enable advanced preparations for these events.

The most recent El Niño that occurred was called the "Super" El Niño. This event took place from 2015–2016; however, previous "Super" El Niños occurred in 1982–1983 and 1997–1998.

The 1982–1983 El Niño was arguably the most devastating, mainly due to the fact that the trade winds not only collapsed but actually reserved. It caused weather-related disasters on almost every continent. During this El Niño, droughts, dust storms, and fires affected areas of Australia, Africa, and Indonesia. Peru had the heaviest rainfall in recorded history, with 3.4 m (11 ft) in some areas, with rivers carrying 1000 times their normal flow. This event has also been attributed to above-normal temperatures in Alaska, Canada, and the eastern United States

(Gannon, 1987). Numerous deaths and billions of dollars in damage were caused by the 1982–1983 El Niño.

The 1997–1998 El Niño was also record breaking, bringing unusual extremes to parts of the globe. Severe weather, including flooding, tornadoes, and ice storms, occurred throughout the United States. January and February of 1998 were the wettest and warmest first two months for the United States in 104 years, according to the NOAA (Thompson, 2015). Temperatures were recorded above normal on the west coast of South America and sea surface temperatures throughout the equatorial east-central Pacific increased to near 29 °C during February and March of 1998, a maxima above +4 °C. Droughts occurred in the Western Pacific Islands, Indonesia, Mexico, and Central America during this 1997–1998 El Niño event ("1997–1998 El Niño," n.d.).

Both the 1982–1983 and 1997–1998 El Niño events were significant enough to have a global impact on the climate to create records and anomalies. The most recent "Super" El Niño phenomenon occurred in 2015–2016 and was one of the strongest in the past 70 years. This event influenced the United Nations to declare 2015 the hottest year since record keeping began. Even though temperatures would have been high without the El Niño, it is possible that 8–10% of the warming was attributable to the event (Cho, 2016). "The Oceanic Niño Index, the three-month-average sea surface temperature departure from the long-term normal in one region of the Pacific Ocean, is the primary number we use to measure the ocean part of El Niño, and that value for November–January is 2.3 °C, tied with the same period in 1997–1998. There are other areas of the ocean that we watch, though, including the eastern Pacific (warmer in 1997/98) and the western Pacific (warmer in 2015/16)" (Becker, 2016).

12.5.6 Extreme Tropical Cyclones

Tropical cyclones are common and devastating at times. They are categorized on scales that classify their strength, providing a warning to society. Tropical cyclones are measured on a tropical cyclone scale that depends on the location of the storm. Whether the storm is located in the Pacific, Indian, or Atlantic Ocean, the cyclone classifications are based upon sustained winds. Although the category strengths vary, the overall wind range remains the same from less than 32 kts (37 mph), to greater than 137 kts (158 mph). However, when discussing the most extreme tropical cyclone on record, it needs to be clarified as to what category of extreme we are referring to. As a society we tend to describe extreme as the damage or devastation that occurs opposed to the record-breaking weather elements and atmospheric processes in and of itself.

An intense tropical cyclone is also dependent upon the pressure on the storm. The lowest pressures observed on earth have been recorded during tornadoes. These small-scale cyclonic events are able to form very low-pressure readings in the creation of a vortex. Tropical cyclones, however, are the second form of low pressures observed on earth. The lower the atmospheric pressure, the more intense the storm.

In 2015, Hurricane Patricia encompassed both attributes, having extremely low pressure and extremely high wind speeds. Located in the eastern Pacific, Hurricane Patricia was a late-season major hurricane that intensified at a very rapid rate. Hurricane Patricia quickly became the strongest hurricane on record to date in the National Hurricane Center's area of responsibility in the western hemisphere. According to the NOAA, Hurricane Patricia recorded a central low pressure of 872 mbar, the lowest ever recorded for an Atlantic and eastern North Pacific basin storm, exceeding Hurricane Wilma in 2005, a previous record in the western hemisphere, by nearly 10 mbar. The rate at which this storm developed into a category 5 hurricane was unprecedented. Over the 24-hour period starting at 0600 UTC, 23 October the pressure fell by nearly 95 mbar and wind increased by nearly 172 mph (150 kts). A tropical wave moving toward the Caribbean Sea interacted with a tropical disturbance that was in the eastern

Pacific on 16 October, and began the formation of this storm. This elongated area of low pressure formed and deep convection increased due to possible favorable conditions associated with a strong oscillation moving eastward across the eastern Pacific and warmer waters due to a strong El Niño (Kimberlain, Blake, and Cangialosi, 2016). On 18 October, the system's cloud pattern began to circulate and become more defined. The system became organized enough to be classified as a tropical depression when it was located to the south-southeast of Salina Cruz, Mexico where a mid-level ridge in the Gulf of Mexico steered the depression west. This was when the storm moved into more favorable conditions that caused rapid intensification. The storm intensified significantly from a tropical depression to a hurricane from 21 to 22 October. The hurricane was categorized as a category 5 hurricane and made landfall in Jalisco, Mexico on 23 October. Although the storm weakened slightly shortly before landfall, Patricia hit a maximum wind speed of almost 200 mph (174 kts) with gusts reaching almost 215 mph (185 kts), only about 15 mph less than the strongest tropical cyclone wind speed on record. The wind speeds were equivalent to a high-end EF4 tornado with a 15-mile-wide span hitting the coast of Mexico. Many scientists were not surprised that this storm occurred during an El Niño period, with a set of atmospheric and oceanic conditions over the Pacific Ocean being ideal to intensify this storm (Kimberlain, Blake, and Cangialosi, 2016).

Another unusual tropical event that took place and broke records occurred in the first week of November 2015. Tropical cyclone Chapala made its way across the Arabian Sea on 1 November. At the peak of intensity the cyclone was equivalent to a category 4 hurricane, with a pressure reading of 922 mbar. This storm brought approximately 61 cm (24 in) of rain to parts of Yemen and its island Socotra, which is more than some of these areas see in seven years (Leister, 2015). Then on 8 November, tropical cyclone Megh made a direct hit on Socotra and then made landfall on Yemen on 10 November.

Cyclone Megh bordered a category 3 equivalent on the Saffir–Simpson hurricane wind scale. In a week's time, two storms both unprecedented in the historical record for this area struck, causing massive devastation. The storms that hit the Indian Ocean basin, which includes the Arabian Sea, are commonly known to be of less intensity, with wind speeds not equivalent to a hurricane on the Saffir–Simpson hurricane wind scale. Before tropical cyclone Chapala, there had been no record of a category 4 hurricane ever tracking as far south as the Arabian Sea (Erdman, 2015). It is not that tropical cyclones are rare for this area, because they actually form at least one or two times each year in the Arabian Sea. The rarity comes from the strength, location of the direct hit, and timespan between consecutive occurrences of Chapala and Megh. The timespan in which these two storms formed is remarkable. "According to NOAA's historical hurricane tracks database, only once before on record have a pair of cyclones tracked within 200 nautical miles of Socotra Island within the same season. Cyclone Twelve passed over the island on 24 October 1972. Less than one month later, Cyclone Thirteen then weakened southeast of Socotra on 21 November 1972. So, it's safe to say back-to-back cyclones affecting Socotra Island within a week's time is unprecedented in the historical record" (Erdman, 2015).

Tropical cyclones occur in cyclical patterns and during certain seasons, dependent on their global location. Speaking in accordance to the known written record of tropical cyclones, it is rare that one forms outside of its season. It is rare because of the atmospheric conditions that are associated with each season. The conditions need to be favorable in order for the formation of a tropical cyclone to begin. In January 2016, Hurricane Pali formed, becoming the earliest central Pacific Ocean hurricane on record. Hurricane Pali strengthened to a category 2 hurricane, and took an unusual track near the equator, which is rarely seen in the central Pacific. "El Niño played a role in the formation of this unusual tropical cyclone. According to

the report issued Thursday 7 January by the Central Pacific Hurricane Center: "This low-latitude out-of-season system has tapped into significant directional shear of the low-level winds, with an El Niño related westerly wind burst south of the system, and prevailing easterly trade winds to the north providing the large scale conditions conducive for development" ("Hurricane Pali Recap, Earliest Central Pacific Hurricane on Record," 2016). The sea surface temperatures for the formation of the storm were estimated at 85.1°F (29.5°C), which was enough to make the area of disturbance an area of low pressure, resulting in a coalescence of a tropical depression and then a hurricane. Another record of this nature had been set by Hurricane Alex, in January of 2016. It is the second hurricane on record to form in the month of January in the Atlantic basin. The last hurricane that formed previously in the month of January in the Atlantic was in 1938. Hurricane Alex is the strongest January hurricane on record, with sustained wind speeds that reached near 85 mph (74 kts) with higher wind gusts. The pressure on this storm was estimated at its lowest to be 981 mbar ("Hurricane Alex," 2016). The hurricane made landfall in the Azores islands of Portugal. These two record-breaking January storms are unarguably outside the normal spectrum. They can be attributed to a number of climatological anomalies along with cyclical patterns on earth. It is near impossible to be certain as to a single factor that led to the formation and maturity of these two storms in the month of January, outside the normal seasons for their development. A tropical cyclone will develop as long as the conditions are favorable for its formation and growth. The responsibility and back tracking of the reasons behind the favorable conditions lead to various scientific explanations being based on, for example, rising temperatures, El Niño contributions, and cyclical changes and patterns of earth.

Tropical cyclones form and mature under specific atmospheric conditions. There are conditions under which growth is enhanced and furthered rapidly, creating an extreme weather event. Recent times have witnessed several extreme events that trump all others in recorded history. On a small scale these events are becoming more extreme as far as the damage they cause and the effects they have on society and the environment. Records have also occurred in scientific terms regarding duration, lowest pressure, and wind speed. Should we begin to expect more extreme events? It is safe to say that society should protect itself from these extreme conditions, due to the fact that population is growing, especially in regions in which extreme events are likely and capable of resulting in the greatest damage.

12.5.7 Droughts

As explained in Chapter 7, Section 7.4, droughts are prolonged periods of abnormally low rainfall totals which are fueled by weather patterns and processes. To classify a drought as extreme, it must exceed the normal length of time of the average drought for that area or region. Also explained in greater detail in Section 7.4, drought has a great effect on the biosphere. Drought is a weather phenomenon that doesn't strike suddenly like a tornado does; however, it creeps up slowly, building as precipitation totals steadily decrease. Data on droughts fixed on a global scale enable research to conclude whether global drought is increasing, decreasing, or remains constant. It can also monitor the intensity, or lack thereof, of these occurring droughts. In recent years global droughts have seemed to last longer and increased in frequency. Drought can be influenced by a number of factors; however, where drought occurs relies on the positioning of ridges and troughs. For example, in the United States, parts of California have recently been experiencing severe drought. This is partly caused by an abnormally strong and blocked high-pressure system over the general region, with a ridge over the western United States. This stagnated positioning of atmospheric circulation has led to an overall increase in high temperatures and an extreme lack of precipitation in the western United States.

This heat and drought is still occurring, even with the help of the most recent "Super" El Niño pattern, when El Niño would normally bring increased precipitation and cooler temperatures to this same area. This El Niño event helped, but it fell well short of ending the overall drought.

With a changing climate, locations of drought are likely to change, causing significant problems for different areas of society. When climate shifts, the positioning of ridges and troughs shift, therefore relocating the areas in which drought would typically occur. A relocation could place drought in areas that rely on rainfall for survival, devastating the area and creating massive loss in all regards. According to the National Science Foundation in April of 2016, atmospheric scientists found trends in southern California temperatures relating to drought. California's highest temperatures are most commonly associated with blocking ridges, in which high pressure disrupts wind patterns, similar to what has been occurring over the past several years. California has developed a nickname to the persistent ridges that appear over the northeastern Pacific Ocean: the "Ridiculously Resilient Ridge," also known as "the RRR." This term appeared in 2013 and has since stuck, due to the current pattern in ridges of high pressure that last for prolonged periods. These long-lasting, high-pressure ridges reduce rainfall significantly in southern California, creating drought conditions. The frequency of these atmospheric circulation patterns that lead to severe drought conditions is appearing to be increasing over time. Research has been conducted to examine the occurrences of extreme high pressure over the same part of the Pacific Ocean. It is very likely that it has been increasing over time and can be linked to environmental changes. Despite the fact that the number of dry patterns has increased, the number of wet atmospheric patterns has not been on the decline. Although this may sound contradictory to the purpose of finding out where the trend is leading, it can be better explained as: drought conditions will become more extreme, instead of average, and potentially lead to precipitation extremes also increasing, instead of remaining average ("California Drought Patterns Becoming More Common," 2016). This again ties into changes in temperature and atmospheric circulation patterns.

Recent studies conducted at Columbia University have shown that mega droughts also occurred in southern California thousands of years ago (Morford, 2015). These studies were conducted by pollen levels in sediment cores observed at the bottom of Lake Elsinore, just southeast of Los Angeles. Analysis showed that times of changes in pollen level, often signifying drought, were associated with warmer oceans that affected the winds along the coast, causing them to blow from the land toward the ocean and decreasing moisture brought to the land. Ocean temperatures off southern California during these mega droughts were similar to recent temperatures, even though the climate conditions were different. Global climate conditions at the time of these studied mega droughts in southern California were cool and wet, much cooler than it is now; therefore, if we experienced this type of drought in current climatic conditions, this mega drought would be amplified significantly (Morford, 2015). There is, of course, like all archeological data, a margin for error and uncertainty; however, this is a basis for research and further digging into past climate drought conditions similar to current trending patterns. This type of research also adds to the ongoing debate and lack of answers when it comes to climate change and its exact influences on the weather, both regionally and globally.

Historic droughts have been recently seen not just in California but also around the globe, from Africa to Brazil, the Caribbean, the Middle East, and beyond. These record droughts can only be compared to as far back as we can date drought conditions; however, these records can signify where conditions are currently trending toward. In South America, parts of Brazil have been experiencing drought since the end of 2014, coming in as the worst drought this country has seen in over 50 years (Johnson, 2015).

This region is prone to drought, but in recent years the drought has become extensive, drying up reserves leading to significant rationing of water. Parts of Asia as well have experienced extreme drought conditions that have not been seen in nearly a century. Droughts in South Africa have forced farmers to cut production and impose restrictions in order to ration water levels. In these conditions, toxins and sewage cannot be flushed out of the rivers, meaning disease will begin to set in, all leading to and enhancing the chances of an increase in famine. Management of drought can become difficult without outside help and societal cooperation (Johnson, 2015). These recent historic droughts have for the most part occurred in regions where they would normally occur. In the past few years, it is their intensity that has increased.

12.5.8 Extreme and Unlikely Snowfalls

Snowfall is directly influenced by temperature and moisture processes. Snow is so reflective that it plays a large role in regulating climate. It reflects incoming sunlight back into space, in turn cooling the earth. If a large snowfall was to cover a large portion of the earth, the reflectivity of the snow would cause a significant lack of absorption of sunlight. This lack of absorption would cause earth's climate to change drastically, resulting in a ripple effect of colder temperatures and overall atmospheric and meteorological changes. On a smaller scale unusual and large snowfall events can also be harmful to the environment and society if snowmelt leads to springtime flooding.

Extreme snowfall can be categorized by not only the amount of snowfall but also the rate at which it falls, or the length of time that the snow is falling. In recent years, global snowfall totals have on average decreased from the anomaly according to the NOAA National Centers for Environmental Information. According to the National Snow and Ice Data Center, in March 2016, the northern hemisphere was deviating by −7.02% from its standard, and the southern hemisphere deviating by +5.44%. The average that these anomalies are based upon is from a 1981–2010 average. While the northern hemisphere appears to be decreasing in sea ice, the southern hemisphere has been increasing ("NOAA National Centers for Environmental Information, State of the Climate: Global Snow and Ice for March 2016," 2016). The southern hemisphere, however, has different characteristics than the northern hemisphere, making it difficult to truly compare the two anomalies. Snow cover extent is sensitive to both regional temperatures and precipitation patterns across all latitudes and longitudes ("Southern Hemisphere Snow Cover Extent," n.d.).

While this is the global average, regionally specific snow events lead to different results. Large amounts of snowfall in higher altitude occur regularly. Large amounts of snow in short periods of time typically fall within higher altitudes on mountaintops. In March of 2015, Capracotta, Italy was a slight exception to the normal snow amounts as it saw 256 cm (100.8 in) of snow during an 18-hour period, according to MeteoWeb (Almasy, 2015). This area is known for its dramatic weather due to its location, which is approximately three hours east of Rome, but this event is debatably record breaking.

March of 2016 had its very own unprecedented snowfall event in the Mexico Sierra Madre Occidental Mountains. A low pressure in the upper atmosphere near California moved south toward Mexico on 11 March, where it was nearly detached from the main jet stream and stalled. This movement brought snowfall to parts of Mexico that very rarely experience it (Dolce, 2016). The low pressure that formed in northern Mexico was strong and unprecedentedly cold for this area, which resulted in snowfall throughout the Mexico Sierra Madre Occidental Mountains.

Mexico's Guadalajara, a city lower in latitude than Miami, Florida, experienced its first snow since 1997, which happened to be part of another strong El Niño setup, just like the 2016 storm (Matz, 2016).

Chapter Summary

Extreme weather is a very important topic in today's world. There are numerous theories and debates on what exactly causes extreme weather, and how we should prepare for extreme weather in times of climate change. Climate change has been measured and researched with state-of-the-art technology. This research shows that the earth is changing in regards to climate. What research doesn't answer is how that specifically relates to extreme weather. There is evidence that more severe and record-breaking events have taken place across the globe, with droughts, tropical cyclones, snowstorms, and extreme heat never before seen and occurring in unprecedented locations around the world. Evidence is lacking on the direct relationship between climate change and extreme weather events. There is an obvious relation, as climate change will likely alter global temperatures, jet stream positioning and strengths, and ocean current direction and strength. What isn't obvious is just how these changes will affect weather and when they will take place, where they will occur, and for how long. Another area of weakness when it comes to extreme weather knowledge is in its actual definition. We can attempt to measure extreme weather by placing scientific thresholds through statistics or by classifying events as "extreme" based on damage or casualties. This gets an overview out to the public and other scientists, but a concrete understanding on extreme weather occurrences is still lacking. More research is needed to better understand extreme weather as a sole element in meteorology, and even more research is needed to understand how it will change in rate of occurrence during climate changes. Mankind has to understand that weather and climate are ever changing, and that the future has a great unknown in store. Advancements in extreme weather prediction will be essential to prevention of loss of life and damage to the biosphere. More research and more powerful tools will enhance and project these ideas and motions forward.

References

Book References

Ahrens, C.D. (2007) *Meteorology Today:*
An Introduction to Weather, Climate, and the
Environment, Eighth Edition, Thomson
Brooks/Cole, Thomson Higher Education,
Belmont, CA.

Cullen, H. (2010) *The Weather of the Future*,
HarperCollins Publishers, New York.

Fry, J.L., Graf, H., Grotjahn, R., *et al.* (2010)
The Encyclopedia of Weather and Climate
Change: A Complete Visual Guide,
University of California Press. Berkeley, CA.

Journal/Report References

"California Drought Patterns Becoming More
Common." (2016) Press Release 16-038. 1 April
2016. National Science Foundation, http://
www.nsf.gov/news/news_summ.jsp?cntn_
id=137911. 16 July 2016.

Easterling, D.R. (2007) Observed changes in the
global distribution of daily temperatures and
precipitation extremes. Cambridge University
Press 2008 and Cambridge University Press,
2009. Abstract.

"Extreme Weather." (2014) National Climate
Assessment, http://nca2014.globalchange.gov/
highlights/report-findings/extreme-weather.
13 July 2016.

Gannon, R. (1987) Solving the puzzle of El Niño:
Is this the key to long-range weather
forecasting? *Popular Science*. 1 September
1986, Bonnier Corporation.

Gornitz, V. (2007) Sea Level Rise: After the
Ice Melted and Today. Science Briefs.
NASA. Goddard Institute for Space Studies.

"Hurricane Alex." (2016) Bulletin: Hurricane Alex
Advisory Number 4. NWS. National Hurricane
Center, http://www.nhc.noaa.gov/
archive/2016/al01/al012016.public.004.shtml.
15 July 2016.

Kimberlain, T.B., Blake, E.S., and Cangialosi, J.P.
(2016) Hurricane Patricia. National Hurricane
Center Tropical Cyclone Report.
NOAA. NWS.

Kohyama, T. and Wallace, J. (2016) Rainfall
variations induced by the lunar gravitational
atmospheric tide and their implications for the
relationship between tropical rainfall and
humidity, *Geophysical Research Letters,*
An Agu Journal.

NOAA National Centers for Environmental
Information, State of the Climate: Global
Analysis for March 2016 (2016), retrieved on
2 August 2016, http://www.ncdc.noaa.gov/sotc/
global/201603.

Stephenson, D.B. (2008) Definition, diagnosis,
and origin of extreme weather and climate
events. In *Climate Extremes and Society*
(eds H.F. Diaz and R.J. Murname), Cambridge,
Cambridge University Press.

Website References

"1997–1998 El Niño." (n.d.) *Department of*
Atmospheric Sciences, University of Illinois,
http://ww2010.atmos.uiuc.edu/(Gh)/guides/
mtr/eln/rcnt.rxml, accessed 14 July 2016.

"A Record from the Deep: Fossil Chemistry." (n.d.)
Earth Observatory, NASA, http://
earthobservatory.nasa.gov/Features/
Paleoclimatology_SedimentCores/
paleoclimatology_sediment_cores_2.php,
accessed 14 July 2016.

Almasy, S. (2015) *Snow Place Like This Village*
When It Comes to One-Day Accumulation,
CNN. 11 March 2015, http://www.cnn.
com/2015/03/10/europe/italy-possible-snow-
record/, accessed 10 August 2016.

Becker, E. (2016) *February 2016 El Niño Update:*
Q&A… and some Thursday-morning quarter
backing, NOAA Climate, https://www.climate.
gov/news-features/blogs/enso/february-2016-
el-ni%C3%B1o-update-q-a%E2%80%A6and-
some-thursday-morning-quarterbacking,
accessed 15 July 2016.

Cho, R. (2016) *El Niño and Global Warming:*
What's the Connection, Phys Org. 3 February
2016, Earth Institute, Columbia University,
http://phys.org/news/2016-02-el-nino-global-
warmingwhat.html, accessed 15 July 2016.

"Convection, Circulation, and Deflection of Air."
(2016) *Atmosphere*, Encyclopedia Britannica,

https://www.britannica.com/science/atmosphere/Stratosphere-and-mesosphere#ref952920, accessed 28 July 2016.

Dolce, C. (2016) *Extreme March Weather Pattern Yields Snow in Mexico, Historic Flooding in South and Record Northeast Heat*, The Weather Channel. 12 March 2016, https://weather.com/news/weather/news/extremes-mexico-snow-south-flooding-northeast-heat, accessed 10 August 2016.

"Drought.gov." (2016) *Palmer Drought Severity Index*, https://www.drought.gov/drought/content/products-current-drought-and-monitoring-drought-indicators/palmer-drought-severity-index, accessed 12 July 2016.

Erdman, J. (2015) *Cyclone Megh Makes Direct Hit on Socotra: Second Final Landfall in Yemen*, The Weather Channel. Hurricane News. 9 November 2015, https://weather.com/storms/hurricane/news/cyclone-megh-five-yemen-socotra-somalia-arabian-peninsula, accessed 15 July 2015.

"Hurricane Pali Recap, Earliest Central Pacific Hurricane on Record." (2016) *Hurricane News*, The Weather Channel. 14 January 2016, https://weather.com/storms/hurricane/news/tropical-depression-one-c-pali-central-pacific, accessed 15 July 2016.

"India Records Its Hottest Day Ever." (2016) *India: BBC News*, 20 May 2016, http://www.bbc.com/news/world-asia-india-36339523, accessed 15 July 2016.

Johnson, T. (2015) *California Isn't Alone: Historic Droughts Happening Around the World*, The Weather Channel. Environment. 28 July 2015, https://weather.com/science/environment/news/california-historic-drought-world-brazil-africa-korea, accessed 17 July 2016.

"June Marks 14 Consecutive Months of Record Heat for the Globe." (2016) *NOAA*, http://www.noaa.gov/june-marks-14-consecutive-months-record-heat-globe, accessed 2 August 2016.

Leister, E. (2015) *Several Years' Worth of Rain Falls in Yemen from Chapala*, AccuWeather. 5 November 2015, http://www.accuweather.com/en/weather-news/strengthening-cyclone-chapala/53282115, accessed 15 July 2016.

Matz, C. (2016) *March Storm Brings Rare Snow to Mexico: Feet of Rain to the Southeast*, Fox Television Station LLC. 12 March 2016, http://www.fox9.com/news/106195051-story, accessed 10 August 2016.

Morford, S. (2015) *Ancient Pollen Points to Mega-Droughts in California Thousands of Years Ago*, State of the Planet. Earth Institute, Columbia University. 22 September 2015, http://blogs.ei.columbia.edu/2015/09/22/ancient-pollen-points-to-mega-droughts-in-california-thousands-of-years-ago/, accessed 17 July 2016.

"NOAA National Centers for Environmental Information, State of the Climate: Global Snow and Ice for March 2016." (2016) *NOAA*, http://www.ncdc.noaa.gov/sotc/global-snow/201603, accessed 17 May 2016

"Ocean Circulation & Climate." (2002) *Climate Change*. University of California, San Diego, http://earthguide.ucsd.edu/virtualmuseum/climatechange1/10_1.shtml, accessed 31 July 20016.

"Ocean Explorer." (2013) *NOAA*, http://oceanexplorer.noaa.gov/facts/currents.html, accessed 31 July 2016.

Rosenberg, M. (n.d.) *The Next Ice Age: Climate Change, and Global Issues*. About Education, http://geography.about.com/od/globalproblemsandissues/a/nexticeage.htm, accessed 31 July 2016.

Schiele, E. (n.d.) *Ocean Conveyor Belt Impact*, Ocean Motion and Surface Currents, NASA, http://oceanmotion.org/html/impact/conveyor.htm, accessed 31 July 2016.

Schirber, M. (2005) *Global Warming Makes Sea Less Salty*, Live Science, 29 June 2005, http://www.livescience.com/3883-global-warming-sea-salty.html, accessed 13 July 2016.

"Southern Hemisphere Snow Cover Extent." (n.d.) *National Centers for Environmental Information*, NOAA, https://www.ncdc.noaa.gov/monitoring-references/dyk/sh-snowcover, accessed 10 August 2016.

Sumner, T. (2016) *Atmospheric Tides Alter Rainfall Rate*, Science News, 19 January 2016, https://www.sciencenews.org/article/atmospheric-tides-alter-rainfall-rate, accessed 13 July 2016.

Thompson, A. (2015) *How This El Niño Is and Isn't Like 1997*, Climate Central's Science Journalists and Content Partners. 23 July 2015, http://www.climatecentral.org/news/comparing-el-nino-to-1997-19278, accessed 14 July 2016.

"US Climate Extremes Index (CEI): Introduction." (n.d.) *NOAA*, http://www.ncdc.noaa.gov/extremes/cei/introduction, accessed 14 July 2016.

Figure References – In Order of Appearance

"Global Analysis – June 2016." (2016) NOAA National Centers for Environmental Information, https://www.ncdc.noaa.gov/sotc/global/201606, accessed 15 July 2016.

"US Climate Extremes Index (CEI): Graph." (2016) https://www.ncdc.noaa.gov/extremes/cei/graph. 12 August 2016.

"Hydrological Information." (2012) Delaware River Basin Commission. nj.gov/drbc/hydrological/, accessed 12 August 2016.

Small, C., V. Gornitz, and J.E. Cohen (2000). Coastal hazards and the global distribution of population. *Environ. Geosci. 10 June 2016.*

Henson, B. and Trenberth, K. (2014) El Niño, La Niña & ENSO FAQ. AtmosNews. NCAR. UCAR, https://www2.ucar.edu/news/backgrounders/el-nino-la-nina-enso-faq, accessed 2 August 2016.

13

Additional Meteorological Impacts

Nicole Gallicchio

It is sometimes easy to get consumed by our daily lives and own personal viewpoint that we forget we are a part of a very large universe that can consume earth entirely in the blink of an eye. A somewhat unexplored area is the spatial threat that is posed to earth and its atmosphere. Space weather and wind, meteors, solar flares, and solar winds are all potential dangers that earth can encounter. These threats have a range of possible effects, from the extinction of the biosphere completely to altering our weather and atmosphere in the troposphere. Threats outside earth's atmosphere are not necessarily restricted to space. There is a whole other impending hazard below the surface of the earth. Earth's crust poses a deeper issue than many may think to the weather and atmosphere. The interactions and exchanges of gases between the surface and below it can impact meteorological processes. Influences external to the biosphere and troposphere inherently have the potential to effect climatic changes on a global scale. In order to protect earth and its life forms, not only is it necessary to explore, research, and document data from the surface but also it is essential that we do the same from the deepest parts of the ocean to the farthest reaches of outer space. Knowledge can only increase our awareness and preparations for future threats.

13.1 Space Weather

Earth has to be concerned not only with weather in our living atmosphere but also with conditions beyond that atmosphere. Space weather is "conditions on the Sun and in the solar wind, magnetosphere, ionosphere and thermosphere that can influence the performance and reliability of space-borne and ground-based technological systems and can endanger human life or health" ("Space Weather," 2016). Severe space weather can have significant impacts on earth. It can disrupt such functions as satellite operations, power grids, marine and aviation technology, GPS systems, and communication technology. Space weather is something that humans cannot control, and have a difficult time preparing for and protecting ourselves against. Space weather occurs in the thermosphere, magnetosphere, and ionosphere: the topmost layers of the atmosphere.

13.1.1 Space Weather and Technology

It is common knowledge that the sun plays a major role in earth's weather and atmosphere. There is growing research and data that support the idea that the 11-year solar cycle affects climate and temperatures, with a good example

The Evolution of Meteorology: A Look into the Past, Present, and Future of Weather Forecasting, First Edition.
Kevin Anthony Teague and Nicole Gallicchio.

being that of "Europe's Little Ice Age, when the sun went through several nearly sun spotless cycles from 1645 to 1715" (Zastrow, 2016, p. 26). Increased solar activity can heat the stratosphere, changing the patterns of wind around the globe, it can alter earth's magnetic field, and it can destroy the ozone layer. Models are now in use and being developed to help show the cycles of the sun, and the effects that it will produce on earth. The Whole Atmosphere Community Climate Model (WACCM) is produced by the National Center for Atmospheric Research (NCAR). The latest of these models, WACCM4, is currently being used to study all these factors. While there are some biases within the model, it is hoped that it will contribute to further studies of the solar cycle's impact on weather and climate (Zastrow, 2016, p. 26). There is also a theory that states that cosmic rays can alter cloud structure and formation by creating nucleation sites in the atmosphere which are responsible for the formation of clouds. This in turn leads to cloudier conditions. An increase in cloud cover will have obvious effects on earth's climate and weather ("Space Weather Impacts on Climate," 2016). Other effects of solar activity, such as strong solar storms, have impacts far greater than weather and climate, and are much more immediate. Strong storms may have the ability to destroy earth's power grid. Tom Berger, director of the National Oceanic and Atmospheric Administration's (NOAA) Space Weather Prediction Center, states that a strong solar storm "would be the equivalent to having several hurricanes come ashore across the east coast [of the United States] at the same time, in terms of widespread damage it could cause to the grid" (Wendel, 2015). Currently, the United States has only one satellite that can collect data on the interaction between solar wind and the earth's atmosphere, and this satellite is over 20 years old, with data sometimes taking hours just to download. Berger states that research and national strategies are needed to be able to say in the future that "we are ready" should a strong solar storm hit earth (Wendel, 2015).

Europe is also following suit on the idea of better protection from the sun's activity. Europe is developing a warning network to help protect its citizens from potential disaster. Today's information relies on data gathered from sensors on the ground and in space. Europe's *Proba-2* sun-watching satellite also helps contribute to data collection on solar activity. In 2016, it is expected that the European space weather network will "encompass over 140 separate products providing scientific and pre-operational applications as part of 39 services provided to users" ("Europe Comes Together for Space Weather," 2015). Space weather, like earth's weather, is a fluid field and full of great uncertainty, but current knowledge and research will help limit the uncertainties, and generate a better understanding and forecasting of future solar activity.

13.1.2 Solar Flares and Wind

The sun is the primary driving force of climate with the immense amount of energy that it provides to earth. The sun radiates heat and light, a small fraction of which the earth absorbs making life possible. Without the sun there would be no weather on earth, or life for that matter. Solar energy in not a constant; it has spouts of solar maximums and minimums. The effects solar energy has on earth can be profound. Variations in solar activity can alter earth's rotation, varying the length of day, which in turn forces surface temperature variations. Variations in the weather or seasons are primarily driven by the sun's output. The sun goes through cycles in which sunspots increase and decrease over time (Figure 13.1). These cycles have approximately 11-year spans and are driven by the reversal of the sun's magnetic poles (Phillips, 2013). The more sunspots, the more solar activity, designating a solar maximum. The solar cycle activity is regarded as one of the major factors that determine global temperatures. An increase in sunspots correlates with an increase in global temperature, while the opposite occurs with a decrease in sunspot activity. The current

Figure 13.1 The sun goes through 11-year cycles in which sunspots increase and decrease. Current evidence points toward a decreasing trend in sunspots since 2015, with this trend anticipated to continue through 2019. A decrease in sunspots has been seen to correlate to a decrease in overall global temperatures. *Source*: Courtesy of NOAA Photo Library.

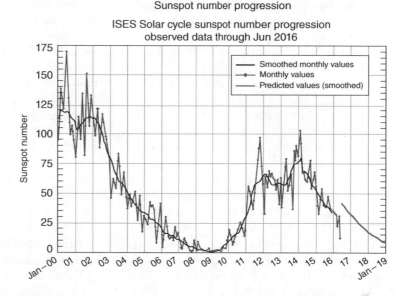

solar maximum peaked in 2013, suggesting that global temperatures would have been peaked at this point; however, there is debate that, like water, the atmosphere is lagged in its absorption of this energy and temperature increase. If this is the case this would place the global temperature peak a few years later. Besides the occurrence of solar flares, there are also coronal mass ejections (CMEs). CMEs occur when magnetic field lines of the sun are contorted and throw solar matter into space. These hot magnetized particles of solar matter are called "plasma." When these large clouds of plasma and magnetic fields move toward earth they may have some kind of impact ("Solar Storm and Space Weather – Frequently Asked Questions," 2016). Many impacts that earth will receive from these types of phenomena will be in the form of technological disturbances that will hinder communication on earth. Disruption of satellite communication and telecommunication is a significant disturbance that poses a potentially great threat to human life. Being ignorant of forthcoming extreme weather or different atmospheric elements creates a scary scenario for society.

13.2 Earthquakes and the Effect on Climate Change

One of earth's naturally occurring phenomena, earthquakes, take place right under our feet, rumbling the ground and shifting the tectonic plates. Earthquakes are a result of a sudden release of energy in the earth's crust that creates waves of energy that break through the crust, known as "seismic waves." Earthquakes create significant damage to society and the environment every year. There are approximately 50 earthquakes that occur every day, and about 20,000 each year. Not all earthquakes are devastating; a large number of them cannot even be felt.

It is uncertain whether earthquakes have an effect on the atmosphere and by default on the weather; however, speculation and research can and should be conducted. Dissecting the workings of an earthquake in connection to the earth's surface can show any potential contribution to climate change. In simple terms, when an earthquake occurs, the earth's surface shakes, altering the landscape and rustling up dirt, dust, and debris. Enough dirt, dust, and debris in the atmosphere can lead to air pollution and blocking out of the sun. A large earthquake would have to occur, creating massive amounts of damage for enough dirt and debris to block out a large portion of the sun's rays to have an effect on the weather. Blocking out the sun would give the world a nuclear-impact-type winter.

The idea of the weather affecting the earth's crust, which in turn triggers an earthquake, is the converse to this. The atmosphere pushing down on the earth's surface and influencing seismic activity has been a topic of interest. It has been widely shown that seismic activity can be triggered by stress changes of barometric pressure over periods of tens of seconds associated with the passing surface wave trains from remote earthquakes (West, Garnero, and Shirzaei, 2015). There have been links to landslides, volcanoes, and earthquakes with sudden drops in pressure due to this pressure drop causing a sudden rise of water and air to shoot upward toward the earth's surface. This rise of water and air reduces friction between tectonic plates, causing a mobilization of the plates, triggering an earthquake. It should be noted that low pressure is not a cause of an earthquake, just one of the many things that can act as a trigger. Earth's tectonic plates are too complex for their movements to be predicted, thus making the prediction of earthquakes and volcanic eruptions severely limited.

13.3 Weather and the Effects on the Human Body

Weather impacting the biosphere, of course, has an influence on the species that live within it. A very controversial debate had among scientists, doctors, non-medical doctors, and society is the effect that weather has on the human body. As climate begins to change, the human body and its reactions to weather may also change along with it. Different temperatures, seasons, pressure, and moisture contents all play pivotal roles on human health. In general, the weather is one of the most frequently introduced topics, often a conversational icebreaker. A large majority of this conversation relates to complaints of aches and pains that are brought about by pressure, temperature changes, and moisture content. It is a safe, neutral topic of conversation because most people will be able to relate to the weather, as it is ubiquitous and affects everyone.

Many argue that this is a psychological, and not physical, effect, while others argue that the changes in weather directly act on the body. The physical body is directly affected by certain meteorological processes, some of which can be fatal. The question of the psychological lies within the aches and pains of joints, muscles, sinuses, and headaches, and their relation with impending inclement weather.

There are several factors that undoubtedly directly impact the human body. One factor on the human body that cannot be denied by any party is the fact that extreme temperatures can be significantly hazardous to and potentially fatal for the body. Maximum and minimum temperatures have substantial impacts on mortality rates, especially in large cities. Threshold temperatures for cities have been determined to protect society. Extreme heat causes the human body to experience numerous medical complications. Hot weather causes the body to overheat and overwork itself in order to try to cool itself down. Such conditions occur as cardiac arrest, asthma, stroke, and dehydration. Extreme cold is not responsible for as many deaths as hot weather. However, it is still dangerous and even deadly. Hypothermia occurs when the body temperature falls below 35 °C, which happens during exposure to cold weather or water. Humidity also has an impact on the body's ability to cool itself by the means of perspiration, leading to heat stress. Low humidity and dry conditions can cause stress on the nasopharynx and trachea, causing difficulty with breathing. These conditions are easily replicated and can be readily reconstructed to demonstrate the impacts on the human body. It is harder to say with any certainty what effects, if any, that meteorological processes such as frontal passages, sunshine, cloud cover or pressure changes have on the human body, mainly because they do not affect people in the same way. The same pressure change could affect one person significantly and another person could have no symptoms at all. A large majority of the population reports having at least once in their lifetime feeling aches and pains in relation to pressure changes, the incoming of rainy/snowy/damp weather, and seasonal changes. Studies have been conducted to try to relate these meteorological processes with physiological changes in the body. This topic has become very subjective, as some scientists have conducted research in pressure chambers showing that the changes in atmospheric pressure are too small to prompt any physical changes in the body. Many debate this topic due to the physical effects people feel on a daily basis in conjunction with weather changes.

It is a difficult to have a steady variable in testing pressure and temperature changes in conjunction with the body, due to the fact that not everyone feels the same changes. However, the majority of our joints are surrounded by a capsule that contains fluid. This capsule is not a closed system, because the fluids can move in and out freely and can regenerate. If you have had an injury to a specific area in which the normal balance of fluid is disrupted, the changes in barometric pressure can cause symptoms in the joint by the expansion and contraction of fluid within the capsule. In times of low atmospheric pressure, the fluid can expand, causing pressure and pain in the joint. This can also be said for sinuses symptoms. The sinuses are cavities; however, they are lined with tissue that produces fluid. This lining can expand or contract with changes to temperature or pressure, creating different fluctuations in sinus pressure and symptoms. This sinus pressure can also be attributed to headaches and vision problems.

There are certain conditions that are intensified by changes in weather conditions. Multiple sclerosis is a condition that is heightened by different weather elements. Cold weather can cause further muscle stiffness and spasms, while very hot and humid weather can also enhance symptoms. Hot atmospheric conditions cause the body's temperature to rise, increasing fatigue and decreasing the passage of nerve impulses, causing a worsening of symptoms. The continent of North America contains the largest percentage of the world's population with multiple sclerosis. There have been studies conducted that prove, among other factors, geography to be a large contributor. There are underlying complex connections between epidemiology and geography ("Who Gets MS? (Epidemiology)," n.d.). One's geography and its effect on the human body would depend on the environment and atmospheric conditions. There is a likely connection between weather and the development of multiple sclerosis. Taking into account that the largest number of people with this disease are located within North America and Europe, these climates need to be compared and researched in order to find a specific trigger ("Multiple Sclerosis on the Map: Who's at Risk?" 2015).

Different places around the globe have higher rates of suicide than others. There is a connection between their climate and the increased rate of depression and suicide. These climates are very wet and overcast. The clouds and significantly overcast sky prohibit sunlight from reaching the earth's surface in these regions. When sunlight goes in through the eyes it allows one's brain to release serotonin, a mood stabilizer that is thought to affect sleep patterns and overall mood. It is thought that a lack of sunlight leads to lower serotonin levels – which in turn leads to depression – along with a lack of vitamin D – which is absorbed by the skin. Another secretion that is affected by the body's absorption of sunlight is melatonin. The fluctuation of melatonin levels can alter the body's sleep rhythms, in turn disrupting the natural physiology of the human body. Mood as well is then affected by seasonal changes, due to the changes in sunlight exposure and duration all over the globe.

Blood in the human body also responds to changes in temperature and pressure. The human body contains a long network of blood vessels and arteries that carry blood throughout the body. As temperature and pressure change, the constriction and expansion of these blood vessels change, causing the blood flow to change in response. Diabetes and blood sugar levels are affected by low atmospheric pressure. Low pressure and cold temperature cause the blood to become more viscous. The viscosity of the blood is the thickness of the blood which correlates to the body's ability to channel things in and out of its cells. Low pressure causing viscous blood means that it will become thicker and flow at a slower rate. In hot weather, fewer heart attacks occur because the blood vessels become larger and the blood can flow more swiftly, allowing the heart to not pump as hard, whereas cold weather constricts the blood vessels and causes high blood pressure, leading to more serious conditions. Unfortunately for diabetics, extremes of hot and cold could have an effect on glucose levels. High temperatures can lead to dehydration in the body, which causes high blood sugar levels. This is because less blood flows though the kidneys, which inhibits the normal function of clearing out extra sugars from the body (Orenstein, 2016). This inability to rid excess sugar causes an increase in glucose levels and potentially extreme medical complications. In low or high temperature, it has an effect on blood viscosity and blood vessel size.

Since ancient times, humans have studied the effects weather has on the human body. Hippocrates in 400 BC studied the effect of climate and the terrain on an individual. He defined four types of climates and categorized behavior and appearance for each climate. Hippocrates suggested which people would be victorious in battle based on their environmental conditions in which they were raised. He strongly believed that climate had a direct effect on behavior and appearance ("TTI Technical Reports Compendium," 2012). Identifying four types of climate and terrain and associating them with

four bodily fluids was arguably his greatest contribution to meteorological science. This shows that even as far back as 400 BC humans had made the connection between weather and influences on the human body. Further research and advancements in technology will lead to further understanding of the effects weather has on each individual body.

Section Summary

Current and future ideologies of weather and climate change will continue to be a global topic of utmost importance. This is partially due to the fact that weather occurs on earth every second of every day, moving in fluid motions through the atmosphere. Paleoclimatic studies of past atmospheric element conditions on earth seek to determine where our climate has been and to aid in where our climate could possibly be heading. Earth's climate has fluctuated from one extreme to the next, from ice ages to warming periods, over its lifetime. Determining thresholds for earth's weather will allow occurrences of extremes to be found and monitored, in turn providing insight into what type of climate we can expect in the future. It is necessary not only to look at outside forces on earth's atmosphere but also to study the effects weather has on the tiniest molecules in the human body. Meteorology and human behavior have coexisted since ancient times. The impacts that weather has on the human body have been felt and studied for a long time; however, further in-depth research in this area is required. This can also be said for space weather, the sun and their effect on meteorology on earth. There is an immense amount of energy both on earth and outside of earth that affects not only weather but also climate change. There are interactions and exchanges of gases between earth's surface, the troposphere, and space, causing a ripple effect when one is altered.

References

Journal/Report References

"TTI Technical Reports Compendium." (2012) *Target Training International, Ltd.*, TTI Success Insights.

West, J.D., Garnero, E.J., and Shirzaei, M. (2015) *Earthquake Weather: Relationships between Barometric Pressure Changes and Seismicity*, Arizona State University.

Zastrow, M. (2016) Model of Solar Cycle's Impact on Climate Gets Upgrade. *EOS*, 97 (3), p. 26.

Website References

"Europe Comes Together for Space Weather." (2015) *Technology Org*, Science & Technology News, http://www.technology.org/2015/11/05/europe-comes-together-for-space-weather/, accessed 4 April 2016.

"Multiple Sclerosis on the Map: Who's at Risk?" (2015) *United Spinal Association*, Spinal Cord Resource Center, http://www.spinalcord.org/multiple-sclerosis-on-the-map-whos-at-risk, accessed 11 August 2016.

Orenstein, B.W. (2016) *How Hot and Cold Weather Affects Your Blood Sugar*, Everyday Health, http://www.everydayhealth.com/type-2-diabetes/living-with/how-weather-affects-your-blood-sugar, accessed 18 August 2016.

Phillips, T. (2013). *The Sun's Magnetic Field is about to Flip*, NASA, https://www.nasa.gov/content/goddard/the-suns-magnetic-field-is-about-to-flip/#.V7xhFvkrIkV, accessed 11 August 2016.

"Solar Storm and Space Weather – Frequently Asked Questions." (2016) *Sun-Earth*, NASA, http://www.nasa.gov/mission_pages/sunearth/spaceweather/index.html, accessed 11 August 2016.

"Space Weather." (2016) *Solar and Heliospheric Observatory*, NASA, http://sohowww.nascom.nasa.gov/spaceweather, accessed 31 July 2016.

"Space Weather Impacts on Climate." (2016) *Space Weather Prediction Center*, NWS & NOAA, http://www.swpc.noaa.gov/impacts/space-weather-impacts-climate, accessed 4 April 2016.

Wendel, J. (2015) *Protecting Earth from Solar Storms*, EOS. Space Science and Physics, https://eos.org/articles/protecting-earth-from-solar-storms, accessed 4 April 2016.

"Who Gets MS? (Epidemiology)." (n.d.) *National Multiple Sclerosis Society*, http://www.nationalmssociety.org/What-is-MS/Who-Gets-MS, accessed 31 July 2016.

Figure References – In Order of Appearance

"Solar Cycle Progression." (2016) *Space Weather Prediction Center*. NOAA. NWS, http://www.swpc.noaa.gov/products/solar-cycle-progression, accessed 4 July 2016.

Section V

The Future Direction of Meteorology

What we still do not know about meteorology is something that we cannot measure. There are constantly new discoveries in the field of meteorology, furthering our knowledge, but also furthering the fact that we can always learn more. We can create better forecasts, warnings, long-term predictions, and global preparation and responses to all types of weather phenomena. The future of weather forecasting is a booming science, and while advances occur within individual nations, it will take a global approach to expand and reach the masses.

14

Weather Technology

Kevin Anthony Teague

Weather technology has grown from original scientific methodology of weather dances and astronomical predictions to the modern era of smartphones, the Internet, and beyond. While weather technology will never hit a plateau, there are some specific areas within weather forecasting and meteorology that are at the forefront at this specific time in terms of need for advancement. Continuing scientific research will need not only advancements in knowledge of the physical elements but also correlating technological advancements to model the performance of the elements.

14.1 Tropical Cyclone Technology

One of the more devastating and powerful meteorological events is the tropical cyclone, with winds of greater than 130 kts (150 mph) at times, storm surges of possibly over 6 m (20 ft), rainfall totals of over 50 cm (20 in), tornadoes, massive devastation, loss of life, and extreme disruption to local and national economies. Due to its intense power, the tropical cyclone is a main focus of meteorological advancements. The threat and devastation of climate change weather scenarios is far reaching. Such examples of these scenarios include the occurrences of

tropical cyclones threatening regions that are unaccustomed to this type of weather phenomenon ("New Report Finds Human-Caused Climate Change Increased the Severity of Many Extreme Events in 2014," 2015). Climate models also indicate that the frequency of powerful tropical cyclones will increase along northeast Australia. In general, the regions with a maximum intensity of cyclones are shifting across the planet, further hammering home the risk for islands like Hawaii, the northeast coast of the United States, and further north along the west Pacific coast of Mexico (Hannam, 2015). In order to better predict cyclonic strength, location, patterns, movement, and likelihood, technological advancements are necessary. Further development in areas such as drone technology, aircraft and sensor technology, better computer initialization data, sea foam detection methods, and improved modeling is needed.

14.1.1 Tropical Cyclone Hunter Technology

Tropical cyclone "hunting" is the main way meteorologists get real-time data of these powerful storms. The more commonly known aircrafts for this type of research are found in the United States, where the National Oceanic and Atmospheric Administration (NOAA) and the US Air Force operate. These pilots and

The Evolution of Meteorology: A Look into the Past, Present, and Future of Weather Forecasting, First Edition.
Kevin Anthony Teague and Nicole Gallicchio.

scientific personnel are designated as "hurricane hunters" (Figure 14.1). When tropical systems, or hurricanes in the Atlantic and Eastern Pacific, become a possible threat to human life and activity, these hurricane hunters become active, flying out and into the storm every 6 to 12 hours. Some data are used for investigative purposes, to gather research on the meteorological and atmospheric structure of tropical cyclones, but also for forecasting the development and movement of the system (Feibel, 2014). One of the more typical aircraft used by the National Hurricane Center (NHC) of the United States is the WC-130J. This aircraft has a crew of five: the pilot, co-pilot, navigator, flight meteorologist, and weather reconnaissance loadmaster, each with specific roles to ensure mission success. Flying into cyclones is an obvious risk, especially when entering through the eyewall of the cyclone. The extreme downdrafts of the eyewall can create the most extreme flying conditions any pilot will ever have to endure. To try to limit this risk as much as possible, flights are done at 3000 m (9800 ft). Both the "hunters" and the aircraft are equipped with measurement technology and tools, transmitting back air pressures, wind speeds, humidity levels, temperatures, visuals of cloud formations, and even radars and sensory technology that pick up a multitude of information and data. Hunters also are equipped with dropsondes, similar to those in weather balloons. These dropsondes are then dropped into various locations throughout the cyclone, picking up data as they fall through the storm toward the surface. Without this technology, most of the information acquired by hunters is at upper levels and altitudes. This brings information downward toward the surface ("Hurricane Hunters," 2015). In fact, the dropsondes have just been upgraded by the National Hurricane Center (NHC) and the National Center for Atmospheric Research (NCAR) in hopes of gathering even more surface data. Added to the dropsondes will be a sensor to measure the water surface temperature at the moment the dropsonde reaches the water. This once-missing piece of data can be looked at as a building block to the structure of the cyclone, as these data at the air–sea interface give scientists a glimpse of what is occurring at the location where energy transfer occurs. This happens to be situated at the most difficult location for any observational tool to be implemented ("Hurricane Scientists Bring a New Wave of Technology to Improve Forecasts," 2016). In the years between 1996 and 2012, there were over 13,000 dropsondes deployed inside 120 different tropical storms and cyclones by the NOAA. It has been reported that dropsonde technology has improved tropical cyclone track forecasting by 32% and intensity forecasting by 20%. Goals in the upcoming years are to combine data and dropsonde technology with other agencies around the world in order to increase the value and quality of the datasets, and to expand the applications and uses of that collected data (Wang, *et al.*, 2015, pp. 970–971).

14.1.2 Sea Foam Detection

A more recent addition to the arsenal of the hunters in hopes of gathering more surface data from the WC-130J is the stepped-frequency microwave radiometer, known also as "smurf" ("Hurricane Hunters," 2015). Smurf is designed to measure the ocean surface winds directly below the airplane, filling a major gap in what was once a missing piece of ocean data information. As the aircraft flies through a storm, the smurf detects the microwave radiation that is naturally emitted from sea foam that is created from the turbulent winds on the water's surface. The computers onboard the aircraft can use this information to calculate wind speed of the cyclone at the surface, a once estimated variable of a cyclone. Wind speed is not the only use of the smurf. The smurf also helps determine rainfall rates within the storm ("Hurricane Hunters," 2015). This way of utilizing the ocean to help forecast can also provide insight to ocean upwelling and overall ocean-surface relationship. Sea foam detection is an important topic for advancements, due to its ability to aid in

Figure 14.1(a) Photo of a hurricane warning *c.*1938. United States Coast Guard aircraft would drop hurricane warnings to the fisherman below out on the open waters. *Source*: Courtesy of NOAA Photo Library.

Figure 14.1(b) Photo of an NOAA's Gulfstream IV-SP jet N49RF in the foreground and the Lockheed WP-3D Orion turboprop N43RF in the background. Both aircraft are members of the large hurricane hunting and weather research fleet of the NWS and NOAA, which also includes the WC-130J. *Source*: Courtesy of NOAA Photo Library.

different surface forecasting, especially out in open waters. Sea foam can help with data initialization and with modeling different atmospheric layers, from the surface to the troposphere.

14.1.3 Drones

Drone technology has become something humans once only envisioned when watching science fiction movies. Now, anyone can go into a store and buy a remote control drone, and off it goes, sometimes with cameras attached, recording bird's-eye views of landscapes far and wide. This same technology, although much more advanced, can and has been used in tropical cyclone research and data collection, often at a much lower expense than many other technologies. One main purpose of drones is imaging. Sensors, cameras, and broadcasting capabilities can be built in with the drones. This technology brings a more accountable and firsthand approach to gathering information

and data. While satellites and radar may do a fantastic job, drones are the best for acquiring data, moving directly where information is needed or missing, and creating a more complete picture for the scientist interpreting the data. Drones can also fly where humans cannot readily or easily get to, such as high terrain, open waters, or flood zones. While most envision drones as somewhat toy-like, they can also be industrially and militarily used, flying at extremely low or high altitudes. All in all, the main advantages of drone technology is that they are lightweight, easy to transport, comparatively low cost, have higher resolutions, can fly at a variety of altitudes, can use camera and video technology to collect and record data, and can get to locations not normally accessible. There are some disadvantages, however, including weight restrictions of instrumental data added to the drone, air space limitations, and wind hazards. When it comes to tropical cyclones, drones can be used in areas around the system in order to obtain more accurate readings of the atmospheric data needed to implement in modeling. Drones are capable of flying into the eye of the storm in which weather conditions are quite calm, allowing for the safe recording of data from a location that a manned aircraft cannot safely get to with options to change altitudes for full measurement of internal atmospheric conditions. This method allows for getting better surface pressure readings, and cloud structure analysis within the eyewall (Harriman and Muhlhausen, 2013, pp. 1–9). Drones also help with aftermath situations, including damage estimating, and search and rescue (there is more on this in Chapter 15, Section 15.3.1). Drones allow for a steady acquisition of data as well, unlike weather balloons, aircraft, or dropping sensors into cyclones (Baxter and Bush, 2014). One example of an unmanned drone actually used for high-altitude hurricane observing in the eastern Pacific is NASA's Global Hawk, which is capable of flying at altitudes of 60,000 feet (18,000 m). This type of drone is capable of flying over the hurricane or tropical cyclone, collecting data similar to that of manned cyclone-hunting aircraft. Increasing the capabilities of unmanned aircraft will limit the risk of those onboard manned aircraft completely, allowing for the 100% safe acquisition of data ("NASA's Global Hawk Drone Aircraft Flies over Frank on the GRIP Hurricane Mission," 2010). Improving the technological capabilities of drones and other unmanned aircraft not only reduces human risk but also will allow us to gather information that before was impossible to get. With real-time cameras on board, weather instruments collecting data, and eventually possibly radar and satellite like technology, meteorologists may be able to get inside a tropical system and see every nook and cranny. This potential for having an extremely detailed understanding of a tropical system will benefit our overall knowledge of cyclones as well as more accurately handling the storm direction and strength forecasting.

14.1.4 Sea Gliders

Another unmanned advancement underway is the sea glider (Figure 14.2). Sea gliders are autonomous water vehicles that move throughout the water collecting data and then returning periodically to surface to transmit the collected information. These gliders are remotely operated and, instead of gathering a vertical profile of a cyclone, dive into the water to gather data from the water itself around the larger regions where cyclone development is common. The most commonly used measurement involves temperature and salinity. The glider profiles the upper ocean several times a day, diving as low as 300 m (1000 ft) below the surface. Data are transmitted as the glider comes back to and breaks the surface ("Hurricane Scientists Bring a New Wave of Technology to Improve Forecasts," 2016). This glider technology is another excellent advancement that gathers real-time data in locations not previously readily accessible.

14.1.5 Data Initialization

Quite possibly, the greatest limitation to tropical cyclone forecasting is where these systems are located. Storm systems over land have millions of data-collection points and methods pouring into computer models, giving extremely accurate measurements for the modeling of storm systems. Tropical cyclones occur over water, sometimes in the middle of oceans, locations where data collection is sparse or even nonexistent. This limited data collection creates an error likelihood within models that meteorologist are continually trying to fix. While forecasting strength and movement of tropical cyclones have become much more accurate over the years, the need for exponential growth remains. This issue is in the area of data initialization. If the initial data placed into a computer model are inaccurate by a minuscule amount, the result will most of the time be thrown out, hence the current inaccuracies with tropical cyclone forecasting. In order to increase data initialization accuracies and successful forecasting of tropical cyclones, more data have to be

Figure 14.2 An example of sea gliders ready to be deployed to profile the upper ocean in and around Puerto Rico. *Source*: Courtesy of NOAA Photo Library.

collected at the surface and throughout the atmosphere of the tropical cyclone itself. For this, the previously mentioned hunters, dropsondes, smurfs, drones, sea gliders, and satellite data are all necessary.

Besides these technological advancements, there are other more specific advancements and alterations to the models themselves to improve the initialized data. With tropical cyclones over the vast open waters, the need for ocean models is of great importance. A problem in the past was the lack of time and resources to update and run the model. The NHC could not keep up with this model, greatly limiting the initialized data. This may now be resolved as other models, such as the Hurricane Weather Research and Forecasting model (HWRF), use these ocean modeling data. Since the HWRF runs the ocean model, that allows the NHC to work collaboratively with the HWRF as needed, rather than having to run the model itself. Physics upgrades will be underway with the models in areas of convection, mixing, vortex initialization, and

more. Bugs will be fixed and new modeling schemes altogether will be implemented into the models. Some of these older schemes were responsible for the discarding of lines of code, creating large discrepancies in projected movement of cyclones and what actually happened. An increase in the horizontal resolution will also be implemented, which will create a better starting point for the models as it becomes more accurate with more initial information. There will be the coupling together of wind–wave–current interactions, resulting in a much better view of variables such as air–sea fluxes of momentum, heat, humidity, and sea spray. There are also upgrades in the effects that land has with cyclone development, such as surface roughness and land surface evaporation efficiency. These improved surface physics parameterization upgrades are continually building on each other and have only become more accurate and reliable, further reducing tropical cyclone track errors. Advancements and upgrades done in 2012 resulted in a 7% reduction to track errors (Ginis and Bender, 2013), and this will only be improved upon with the current and future planned upgrades. These very specific and technical advancements could be considered the behind-the-scenes fixes and advancements that will improve data initialization, and are just as valuable as the advancements discussed in greater detail earlier in this chapter.

14.1.6 Tropical Cyclone Modeling

Accurate and dependable data initialization is not the be all and end all to the modeling of tropical cyclones. The model itself also has to be reliable and accurate. Great changes have been made over the years to the main tropical cyclone weather models. Some of the greatest changes have occurred over the past 10 years and are projected to continue. With the advancements of computer power and modeling capabilities, "global models can accurately simulate the number and intensities of hurricanes typically observed in different regions of the world" ("A Decade of Progress in Projections and Modeling," 2015). The NOAA states, "Computer modeling of hurricanes and the environment within which hurricanes form and intensify is foundational to providing improved daily and seasonal hurricane predictions. On daily timescales, reductions in hurricane track and intensity forecast errors remain a forecast primary goal" ("A Decade of Progress in Projections and Modeling," 2015). For a period of time, the main focus with modeling was improving the track forecast of a cyclone, while strength forecasting seemed to be put aside. While direction and track are extremely important and, as stated in Section 14.1.5, has seen marked improvements through better data initialization, the strength of cyclones is also a vital piece of information, oftentimes contributing to certain track changes and landfall effects. The HWRF and the Geophysical Fluid Dynamics Laboratory (GFDL) model operational hurricane models have started to improve forecasting. These models have an increased resolution, helping them pick up some of the smaller features within a tropical cyclone. These models also include the data collected by other means, such as drones, aircraft, and dropsondes. The HWRF is not just limited to a single run of the model either. Instead, the HWRF can simulate multiple weather events across the globe. The use of the HWRF in the western Pacific is a major improvement to the forecasting of tropical cyclones. The HWRF takes into account variables such as energy fluxes from the ocean to the atmosphere, rain, moisture, and much more ("HWRF: About Us," 2014). The GFDL model has reduced its intensity errors by 40%. Along with that, this model has also improved certain aspects within it to better predict the amount of tropical cyclones and type of activity expected from months to years in advance. These advancements in long-range tropical cyclone forecasting are only set to be expanded and become more accurate as new developments and advancements are added to these forecasting models. One other model that has just been introduced and has not yet been assessed enough to test its true skill level is the North American Multi-Model Ensemble (NMME). This model, created by the Climate Prediction Center, was developed as a new and improved seasonal hurricane prediction tool. The NMME was first used in the 2015 hurricane season, and even though results are still not quantified yet on its success, the advancements within this model are ongoing ("A Decade of Progress in Projections and Modeling," 2015). With all the improvements in tropical cyclone technologies, the main objective is to continue to help prepare those in harm's way, give as much warning as possible, accurately predict strength and location of certain aspects of the storm (be that rainfall, surge, or wind), and generally help save lives and limit the financial and social strain that a large and devastating tropical cyclone can create. Secondary to that, these advancements help further the research and understanding of tropical cyclones, which in turn helps to further the prediction process again ("HWRF: About Us," 2014).

14.2 Coastal Flooding

While tropical cyclones have become a huge area of focus, somewhat related to the cyclone's effects is the threat of coastal flooding. Coastal flooding is not only due to tropical cyclones. The threat of coastal flooding exists around the globe, especially now with the threat of ice melt from the Polar Regions, creating a rise in sea levels along all the coasts in the world. Nearly 40% of the world's population is located on or within miles of a coastline ("WCRP: Home," 2016), and this percentage is only increasing. To further emphasize the threat of coastal flooding, research has shown that coastal regions will see a faster growth in population than landlocked regions, greatly increasing the threat of larger-scale impacts from coastal flooding events (Neumann, *et al.*, 2015). With massive amounts of people living right on the water, coastal flooding becomes of extreme importance, be that in regards to forecasting, evacuation planning, or even in insurance, zoning, and building laws.

14.2.1 Coastal Flood Risk

In order to discuss flood risk along coastal locations there has to be a way to show and quantify risk zones and to be able to transmit that information to a wide audience of different backgrounds and knowledge of the subject. One of the most basic, yet successful, ways is by mapping and modeling a "damage-flooding depth probability exceedance curve for various scenarios over a given planning period and determining the areas under the curves" (Acton, 2013). While many locations fall below that curve, there are some that stand out as greater risks in the future, due to climate as well as population growth. The coastal regions of Africa are among the regions with the most rapid increase in projected population and in return coastal flooding exposure laws (Neumann, *et al.*, 2015). Coastal threats seem to be at some of the highest risks along the eastern coast of Ghana, in the Volta Delta region. This is mainly due to "human interventions, climate change and the resultant rise in sea-levels, increased storm intensity and torrential rainfall" (Acton, 2013). Asian countries, including China, India, Bangladesh, and Vietnam, all are projected to continue to have the greatest number of total and urban population in coastal flood prone areas (Neumann, *et al.*, 2015). In general, most delta cities are to see the influences of climate change and coastal flooding at a greater rate than any other location. Whether the delta is in a city like Hong Kong, southern China's Guangzhou, or Holland's Rotterdam, all are threatened by an increased flooding risk (Francesch-Huidobro, *et al.*, 2015).

The flood risk in England and Wales has called for great changes in flood risk management for the region. Using a quantified national-scale flood risk analysis, an assessment of England and Wales from the year 2030 through 2100 showed up to a "20-fold increase in economic risk." This same analysis showed that the increase is due to once again the threat of climate change, increased precipitation, and sea level rise throughout the United Kingdom (Hall, *et al.*, 2003). In 2013, the Intergovernmental Panel on Climate Change reported how sea level rise would react and differ by the years 2081–2100, based on the increase of different potential pathways of greenhouse gas. This report shows that the lowest increase in height is predicted to be around 26 cm (10 in), and the highest predicted possibility of sea level rise is 82 cm (32 in) (Henson, 2014, p. 146). Many coastal areas around the world will experience devastating flooding if this range holds true over the next 60 years. Now this may not seem like a major issue to the layman, but think of the amount of flooding and damage that occurs when a coastal storm and or a full moon tidal cycle pushes a surge of 0.3–0.9 m (1–3 ft) into coastal areas. This will then be the norm, with any additional tidal flooding or storm surge on top of that. For every foot increase in sea level, it is said that it is equivalent to an increase of one or two categories in the Saffir–Simpson hurricane wind scale as it relates to storm surge. With this theory, even if storms do not continue to grow in intensity and extreme weather doesn't increase, your average storm will be categorically stronger based strictly on the increase in sea level (Sobel, 2014, pp. 265–266).

Coastal flooding risk is mapped and quantified by the US Federal Emergency Management Agency (FEMA). By 2018, new flood maps will be created, and will cover more than 75% of coastal areas. They will replace 30-year-old maps, which are currently not representative of current climate characteristics and population totals. Land shape and structure of coastal regions also have changed over the 30 years, due to coastal erosion, building, and infrastructure projects. Technology used in mapping has also improved greatly, especially when compared to 30 years ago. Changing the mapping of flood-prone locations has an enormous effect on the safety and economy of a region, with influences on insurance policies, zoning regulations, and evacuation zones ("Coastal Flood Maps," 2016). One thing that is in mankind's favor is that the sea level increase, debatably, is growing at a slow rate. One to three feet (0.3–0.9 m) over the next 60 plus years is a slow enough increase that will allow communities and governments to better prepare for this threat. It isn't a threat that will occur overnight. Instead, years of policy changes, laws, overall awareness, and defenses can take place, limiting the impact of sea level rises (Sobel, 2014, pp. 266–267), and possibly even slowing or halting the rise in general.

14.2.2 Coastal Flooding Technology and Preparedness

Besides the results of better mapping and understanding of risk-prone areas, technological advancements in areas of coastal flooding forecasting will further help with preparing for coastal flooding events. Climate change, sea level increase, precipitation, and other weather phenomena have been discussed, but quite possibly the greatest cause of the most severe flooding along coastal regions is from storm surge. Storm surge is a rise in water levels along the coast usually caused by high winds pushing the water into and onto itself, increasing its

height. Shallow water areas and times of high tide or full moons often add to the height of a surge event. These added variables make forecasting the exact time of a surge very important. A change of a few hours can have a drastic effect on the intensity of the storm surge, such as whether it falls at low tide or high tide. Measuring storm surge can be done through ocean models, taking into account water depth, land shape, embankment of the shoreline on land and just off the coast, wind speed of the weather event, and timing of the event. While these areas are becoming better modeled and understood, technology directly measuring surge height and potential has been implemented and is set to expand. This advancement focuses on gauge and sensor technology. Since the destruction of Hurricane Sandy along the east coast of the United States, the United States Geological Survey (USGS) has taken a strong lead in understanding and documenting the height, extent, and timing of storm surge events. Currently in place along the east coast of the United States are real-time tide gauges set to gather data during hurricanes and coastal storms. The USGS constructed the Surge, Wave, and Tide Hydrodynamics (SWaTH) Network from Maine, southward to North Carolina, consisting of 71 real-time telemetered tide gauges, 61 rapidly deployable gauges, and up to 555 temporary mobile storm-tide sensors (Figure 14.3). These mobile sensors are able to be deployed in advance of storms and in locations of expected surge zones. This is a very new and rapidly growing method of storm surge data collection ("Storm Surge and Coastal Inundation," n.d.). All the information collected from this network is stored nationally, and displayed through a mapping application, allowing other departments and agencies access to the information to help with various planning and safety discussions for coastal locations ("USGS Surge, Wave, and Tide Hydrodynamics (SWaTH) Network," 2016). Louisiana's Lake Pontchartrain saw a large storm surge in 2012 from Hurricane Isaac. In response to that event, the National Weather Service (NWS) and NOAA's Center for

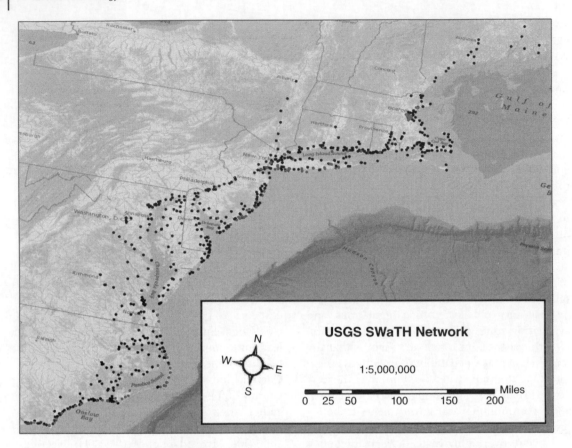

Figure 14.3 A collaborative network of Surge, Wave, and Tide Hydrodynamics (SWaTH) along the east coast of the United States, implemented after Hurricane Sandy. The network consists of 71 existing and new flood-hardened, real-time telemetered tide gauges, as well as 61 real-time gauges, and up to 555 temporary storm-tide sensors. *Source*: Image courtesy of US Geological Survey, Department of the Interior/USGS.

Operational Oceanographic Products and Services (CO-OPS) has installed new real-time data water level stations with microwave sensor technology and leveling surveys to ensure accuracy (Figure 14.4). These stations are built to withstand 110 mph winds and multiple storm events. Their goal is to continuously monitor and collect measurements to show data during calm periods as well as before, during, and after storm events ("New NOAA Lake Pontchartrain Sensors to Provide Better Evacuation Planning, Storm Surge Data," 2015). Water level stations are created to be in a fixed location, allowing for real-time data collection ("Storm Surge and Coastal Inundation," n.d.).

Sometimes a simple database of information can be very useful in understanding the behavior of weather events. In regards to storm surge, there seems to be a real lack of quality databases of surge events. One of the main databases recently created is SURGEDAT for the Gulf Coast of the United States. SURGEDAT collects information as far back as 1880 on over 190 surge events, measuring peek surge heights (Acton, 2013).

Very different, but with some similar consequences is the threat of tsunamis. Although rare in occurrence, they cause some of the most dramatic and life-shattering devastation, as seen with the Indian Ocean tsunami of 2004. Tsunami detection around the world has advanced

Figure 14.4 After Hurricane Isaac in 2012 there was a push for the development and installation of new water level stations. This station uses microwave sensor technology, providing real-time data on water levels and storm surge forecasting. *Source*: Courtesy of NOAA Photo Library.

greatly in response to the 2004 tsunami, through warning systems, sirens, and education. Advancements are with the global network of 60 deep ocean assessment and reporting of tsunami buoy stations (DART) in the Pacific, Indian, and Atlantic oceans. Along with that, 188 water level stations have been implemented ("Advancements in Detection, Forecasting and Safety Result from 2004 Indian Ocean Tsunami," 2014). Water level stations are used to monitor sudden increases and decreases of water depths.

In regards to overall warning systems and public knowledge of flood threats, the public's overall understanding of the potential risks and dangers of flooding needs to be improved. The perfect most recent example of the lack of communication between weather departments and the public was with Hurricane Sandy along the mid-Atlantic coast of the United States. As Hurricane Sandy drew closer to the coast, the storm, according to scientific standards, lost its tropical cyclone characteristics and was labeled an extratropical storm. Now this text won't go into the scientific differences of these two types of weather features, but what is important to understand is how that simple change in scientific terminology created a misunderstanding by the public to what was in store for them. Hurricane warnings were dropped shortly before landfall in New Jersey and its surrounding areas, confusing the public, all the while destruction was wrought through winds, rain, and surge. The fact that the storm wasn't technically a hurricane didn't really matter in the main scheme of things, but the NWS and the NHC decided to cancel the warnings. The guidelines created by the NWS are a little less exacting now, with newer policies in place simplifying warnings and information, even if hurricanes transition to extratropical systems. A better understanding of just how important specific wording is to the public, insurance companies, and emergency management could have greatly benefited and protected those affected by this storm. More specifically related to storm surge, the forecast for this hurricane was too little too late. While it was a relatively accurate forecast ultimately, if the forecasting of a 1.8–3.4 m (6–11 ft) storm surge in New York City had come sooner to the public and governmental agencies, a better plan and preparation could have been taken, possibly saving more lives and property. Evacuation orders could have been implemented earlier, helping those without transportation to have longer to take action, as well as allowing nursing homes and hospitals to have more time to transfer patients and personnel to different locations. Communication between those who issue warnings and those who receive them are improving all of the time, and new and better ideas are coming out each year (Sobel, 2014, pp. 263–264).

14.3 Extreme Weather

Extreme weather events – such as tropical cyclones, coastal flooding, blizzards, drought, extreme heat and cold, wind, tornadoes, and hail storms – have been at the center of weather-themed conversations. Adding to our knowledge of extreme events, as touched on in previous chapters, social media plays a huge role as well, allowing firsthand up-to-the-minute accounts of extreme weather events across the world. The recent technology boom of social media, as well as other meteorological advancements, has brought great attention to the subject of extreme weather. Whether or not all social media is help-ful in increasing our knowledge of and prepar-edness for extreme weather, technology growth in this area is undoubtedly of vital importance. No location across the world is free from the dangers of an extreme weather event ("New Report Finds Human-Caused Climate Change Increased the Severity of Many Extreme Events in 2014," 2015). Appendix Figure G shows the geographical view of the Natural Loss Events Worldwide in 2015. Globally, a lot is being done to address the issues we all face with the risk of extreme weather. Besides the global approach to the issue, there are more small-scale and local-ized technological advancements underway or being planned to help prepare individuals for severe weather, and to help prevent loss of life and damage to property.

14.3.1 Forecasting and Nowcasting

To strive toward the most accurate forecast, meteorologists have to take into account many variables, as we now know. Furthermore, they have to rely on the tools that they have along with their own knowledge and experience to produce the best forecast at that moment.

Shorter-term forecasts have a greater risk of the meteorologist being called out on inaccuracies, as it doesn't look very good at all should a fore-caster have to adjust a forecast greatly while the storm is already underway. An example of this is the potential catastrophic blizzard for New York City in 2015. At the time, the governor of New York City, Andrew Cuomo, decided to take extreme action ahead of the storm to help keep the citizens of New York and New York City safe. The most dramatic action was the closing of the New York subway, for the first time in its 110-year history at the time. While the forecast blizzard of 75–100 cm (30–40 in) of snow in areas did occur, it did not occur where it was originally forecast. The slightest shift of a mere 15–30 km (9–18 miles) resulted in a handful of inches across the city. This led to a great public outcry against the governor's actions, and against the skill of the NWS meteorologists. What the public doesn't realize is that in most cases weather forecasting is an imprecise sci-ence that has to take into account millions upon millions of variables. Recent advancements in meteorology and forecasting have created their own monster; with forecasting becoming much more accurate and over longer periods, these advancements are then taken for granted, leav-ing any larger-scale error all the more glaring. If you enlarge the forecast map up to the size of the world, or even just the size of a country, being off by 10 to 20 miles is an extremely good accomplishment, but when you break it down to localized areas and specific details for those localized areas then the risk of an inaccurate forecast expands greatly. To decrease the risk of inaccuracies, forecasters have been relying more heavily on a form of forecasting called "now-casting," where a meteorologist keeps an eye on the present conditions and short-term future predictions and tendencies. Nowcasting helps pinpoint much more accurately the specific details of an extreme weather event, explaining where certain banding features may set up in a tropical cyclone or blizzard, where thunder-storms are most severe, and where Doppler radar show possible rotation indicating tornadic activity. Nowcasting has become the frontline in the battle with extreme weather. Increasing resources and research for nowcasting may help with the shorter-term reactions by leaders and officials ahead of storms, such as the New York governor's decision to shut down the subway. If better nowcasting had been available at the time for the local meteorologists then quite possibly it could have been seen that the more high-resolution models were suggesting a more eastward trend for the storm, and that may have helped the governor in delaying his decisions just long enough to see the weather conditions verified by the models and in real time. The battle between alerting the public ahead of an extreme weather event and the accuracies of the data available for specific locations at the time of decision-making is where a lot of research and technological advancements are planned and occurring. The problem is that perfecting the timing of a forecast is extremely difficult, and no blame is on any meteorologist in the New York City area, or any area where storms behave differently than predicted. Even with many years of advancements and improvements, forecasts may not get to be much better than they are right now. Instead, the theory of localizing the forecast is becoming much more valuable, and that is where nowcasting becomes the main player. New statistical measurements, data, computer power and models, smartphones, radar and satellite, and more all help personalize a prediction and can even break down weather events to the street level of a city (Summer, 2015).

The first role of nowcasting is to collect data and produce a forecast with a precise geographical display, including information from radar, satellite, weather soundings, and surface data ("Thunderstorm Auto Nowcasting: Overview," 2016). The WMO defines nowcasting as: "the detailed description of the current weather along with forecasts obtained by extrapolation for a period of 0 to 6 hours ahead. In this time range it is possible to forecast small features such as individual storms with reasonable accuracy." Nowcasting is "a powerful tool in warning the public of hazardous, high-impact weather

including tropical cyclones, thunderstorms and tornadoes which cause flash floods, lightning strikes and destructive winds" reducing fatalities and casualties, damages, and economic losses ("Nowcasting," n.d.). Nowcasting isn't just for public use either. It is relied upon by airlines and airports, helping in the decision-making of what terminal or runway is best during weather events, where turbulence is most likely while in flight, and where potential hazards may be throughout the flight. Other areas which rely greatly on nowcasting are boating and marine forecasting, power management, offshore oil companies, construction and outdoor industries, farming, and even everyday life. The most often relied upon tool a meteorologist uses during nowcasting is radar data. Being able to pick up on radar returns for precipitation type and intensities, movement of storms or specific weather bands or cells, tornadic rotation, wind speed, etc., a forecaster can better map out to the street level where specific weather characteristics will occur and when. To expand nowcasting beyond six hours, radar is included with satellite imagery and NWP models, such as high-resolution models ("Nowcasting," n.d.).

Nowcasting is not a new technique in forecasting, having been used throughout the twentieth century, but the new technological developments of today make it a valuable tool and resource for meteorologists, allowing a whole new type of weather forecasting. The Internet, social media, smartphones, and video have all helped disperse data and information more freely and rapidly than ever before.

The biggest of those instruments might be the smartphone. Wireless connections and increased data capacity and phone power have created a personal computer to be held in the palm of one's hand, with camera and video capabilities and the ability to share information instantly. GPS technology also allows specific applications on the phone to give the user their exact location. When you combine that with weather applications or social media, you, or the people viewing your pictures or posts can know what is going on at that exact location

(Mass, 2011). Social media, be it Facebook, Twitter, Instagram, etc., has the ability to instantly connect millions of people to a story or topic. This plays a big role in storm reporting, and preparation prior to weather events. The main thing is to get word out to the public. Once people are aware of a possible weather event, they can decide to take action for themselves and start to pay closer attention to the nowcasting of meteorologists. Smartphone applications also help deliver information to the public. There are applications for users with more knowledge of weather and sciences, which include model data, radars, satellites, weather soundings and more, while others are geared to the layperson, with specific quantifiable data, such as temperatures and percentage changes of precipitation (Mass, 2011).

The NWS itself has taken major steps to improving nowcasting, and making it a main part of its daily routine of weather forecasting delivery. In the past, most updates by the NWS came every six hours, followed by most of the computer models used. Most of the updates that were prepared during these six hours tended to be outdated when it came to imminent weather. Steps have been taken to ensure that nowcasting is more of central to the NWS's work, updating information much more regularly for the short term, and even hourly or more frequently than that when severe weather is present, such as with tornadic activity or hurricane landfalls. Instead of most time being spent on longer-range forecasts, the NWS is making short-term forecasts a priority. This is also the case through televised weather broadcasts. Typically, a brief and broad segment of a few minutes is designated for weather, but during rapidly changing storm developments, meteorologists are seen more frequently describing what is happening to their (usually local) viewers, allowing those in danger to get the most accurate and real-time information. With these improvements and greater value on nowcasting comes another dilemma for the meteorologist. Most long-term forecasting is set on reading models and statistical data and interpreting the

best fit of all information in order to create a forecast. The short-term nature of nowcasting is much more hands on, where the meteorologist has to actively read radar and satellite imagery and interpret information on the spot, all under greater pressure and within a more intense working environment (Mass, 2011).

A major drawback to the expansion of nowcasting is that more and more information is becoming readily available to anyone who wants it. This sometimes creates the sense in some people that they are "meteorologists." This becomes a huge danger and problem, as information then may be dispersed incorrectly to people again and again, creating a false understanding of the short-term weather, possibly creating a deadly situation. The balance of dispersing information to the public rapidly through many different forms and the battle with false distribution of information among the public is one that will have to be addressed by the leading weather agencies and governments. That, along with increasing education and training in nowcasting practices for meteorologists themselves, and increasing funding and research in nowcasting technology, has and should continue to be the main area of focus (Mass, 2011).

14.3.2 Nowcasting Precipitation

One of the more prevalent nowcasting uses is for precipitation amounts and types for localized areas. The global models can show where precipitation will likely occur, and even give a very accurate calculation of the total amount of precipitation. What the global models do not tell you is an exact anything: a storm's exact start and end time, where and when heavier bands of precipitation will occur, how heavy precipitation will be at a specific time and location, etc. These are all areas where nowcasting takes the lead. Globally, some of the spatial resolutions are as much as 25–50 km (15.5–31 miles). This is a great improvement technologically when looked at globally, but locally that is way too large, sometimes larger than the area being focused on itself. Much finer resolutions are required to make any contributions to localized nowcasting. A downscaling technique for global models was successfully tested in 2011. The model tested gave a 24-hour lead-time with accurate results for flooding to occur in the tested location. This technique can be of great value to vulnerable areas and to governmental agencies in order to better prepare those in the flood zone. This can be of enormous value to even less-developed locations and countries, where there is less weather observational data and less high-resolution modeling. This test was conducted from previous data for central Vietnam. Specific algorithms were put together to downscale what the global models predicted for the localized region. At first the algorithms did not show the exact rainfall amounts that occurred, but after taking into account the altitude of the rain gauges that were being used along with other variables, the results were extremely accurate. These data could then be used in flood forecasting when combined with other models for runoff flooding (Do Hoai, Udo, and Mano, 2011). Although this was a study done on old data, the algorithm was used on upcoming storms and verified those as well. This could be a great tool for the short-term forecasting of rain events, especially for those areas with less information available to them.

The development of the Trident system also addresses the issue of precipitation nowcasting. The Trident system is a newer system used to "blend synoptic, mesoscale and storm scale information together in a physically meaningful way, to provide high resolution, timely, quantitative prediction of heavy rainfall and warning guidance products for the potential for flash floods" ("Precipitation Nowcasting," n.d.). The Taiwan Central Weather Bureau, the Beijing Summer Olympics, as well as the NWS have used and tested the system over the past few years. It proved valuable in forecasting thunderstorms near the venues of the Beijing Olympics, as well as flash-flood events and thunderstorms in Taiwan by accurately predicting the initiation, intensity, and duration of the rainfall.

The NCAR and the NWS have tested the Trident system out in the Colorado Front Range of the Rocky Mountains, an area very prone to flash flooding, and in multiple forecast offices throughout the country, providing meteorologists with high-resolution location-specific nowcasting for thunderstorm activity. The main advantages of Trident are that it does not need a lot of computers or systems to run it. It utilizes just one computer workstation. It uses both single- and dual-polarization (dual-pol) radar information, runs in areas that have complex terrain, allows the forecaster to interact with it, and includes a four-dimensional (4D) variational Doppler radar analysis system (VDRAS). This allows for the assimilation of data collected by the radar and placed into a numerical model that then produces a 4D high-resolution model, showing boundary layer winds. The main uses of the system include estimating location and intensity of thunderstorms and rainfall, predicting where convection is likely to occur while monitoring cloud growth, the ability to track gust fronts and wind shear, and an overall better nowcasting toolset for the forecaster ("Precipitation Nowcasting," n.d.).

14.3.3 Nowcasting Tornadoes

Another aspect of nowcasting which has tremendous popularity is the nowcasting of tornado events. Whether that is on television, or with storm chasers, tornado nowcasting is of extreme value and importance. The Severe Storms Research Center (SSRC) at Georgia Tech in the United States has funded numerous experimental technologies to help with the early detection of tornadoes based on low-frequency acoustic waves and a pattern of cloud-to-cloud lightning strikes, which produce specific electrical discharges. SSRC has also been striving for a better 3D display of severe thunderstorms that are able to be understood by most people, a high-resolution model that documents tornado paths and duration, and a more effective tornado-detection algorithm ("Severe Storms Research Center Developing New Technologies to Increase Tornado Warning Time for Georgia Citizens," n.d.). More research has been put into the understanding of the relationship between available potential energy, storm relative helicity, and vertical wind shear quantities. During different seasons in North America, these characteristics play different roles. During colder seasons, vertical wind shear is the main player, while during the warmer seasons, tornado development can be explained by the potential energy and helicity (Cheng, *et al.*, 2016). North America, and especially Tornado Alley in the United States (see Chapter 7, Section 7.3.1), is the most often thought of location on earth that sees this most powerful extreme weather event. This, however, is not the only location in the world that experiences tornadoes. Some of the strongest tornadoes that have occurred have spun up in Australia, New Zealand, East Asia, South Africa, South America, and Europe. Between 1800 and 2014, 9563 tornadoes were reported throughout Europe. Most reports came from the northern, southern, and western regions, and the least in the eastern regions (Antonescu, *et al.*, 2016).

Clearly, tornadoes can occur nearly any place in the world. This, along with their tremendous destructive power, leads a lot of research and technology to nowcasting them in order to protect those in their path. Newer technology for storm chasers allows them to better position themselves while following potential storms, and allows for the collection and quantification of the atmospheric makeup and conditions before, during, and after tornadoes. Radar technology has allowed meteorologists to locate tornadoes based on wind direction and speed changes. This leads to advanced-warning times and better forecasting of potential locations of the storm's path. Also, radar technology can now show debris fields, which again better help a meteorologist figure out the likely location of a tornado. Nowcasting tornado storm cells has already saved countless lives and property. Further developments with radar technology, on-the-ground sensors and storm-chaser technology, and better high-resolution models,

will lead to better lead-times in forecasting tornadic events. With most tornadoes extremely localized and lasting for less than 10 minutes at a time, meteorologists are faced with a huge challenge. Not nearly enough information on localized forecasting of tornadoes has been collected to create a modeling structure that can accurately pinpoint tornadic development, but that is the main goal in the near future. High-resolution models have advanced greatly, showing a great depiction of storm lines and cell activity, but the precise location of tornadoes is still more or less a mystery, leaving much to be desired and needed. This even goes beyond the nowcasting aspect of tornadoes and more toward the mid- and long-range forecasting of tornadoes as well. This is one major area in the field of meteorology that needs to continue to be researched. While convective modeling is becoming stronger and is used along with the high-resolution modeling, there is still no sure-fire way of predicting precisely where a tornado will occur prior to rotation being detected on radar or spotted on land.

14.3.4 Warning Systems

Warning of potential weather-related disasters has seen remarkable improvements over the years and decades. From the unknown to the known, weather events are greatly mapped out with a growing lead-time to prepare for them. Whether this is through weather agencies issuing watches and warnings, or via social media, the Internet, television, or smartphone applications, the word of impending storms spreads much further and faster today than ever before. That, along with weather agencies getting much more accurate data for long-range forecasting, allows a once unthinkable lead-time to weather events. New radar technology, satellite technology, and computer modeling power and capabilities set today's meteorologists leaps and bounds ahead of the meteorologist of just a decade ago.

Warnings can be issued for fog, wind, flooding, marine conditions, thunderstorms, snow or ice, tropical systems, heat, wind chills, and much more. Certain criteria are set for each region of the world by each national weather agency, and when these criteria are forecast or actually occur, watches and warnings are issued. The best time to warn people and prepare them for a weather event is, as Luca Mercalli, president of the Italian Meteorological Society states, when the "sun is shining and not during emergencies"("Researchers Advise Warning Systems and Preparation to Mitigate the Effects of Extreme Weather Events," 2015). The first step to bettering the warning systems for storms for weather events is to improve the computer models and other predicting tools of the meteorologist. A reliable forecast is fundamental in regards to any type of early-warning system for weather events. This reliable and accurate forecast depends on a network of observations from global, national, and regional levels, and from the atmosphere, oceans, and land (Rogers and Tsirkunov, 2013, pp. 18–19). Once forecasting improves, longer lead-times and more accurate predictions of what is to occur will help local agencies take the necessary precautions.

It is also important not to issue too many warnings, as then the value of them decreases. The UK weather service in 2009 and 2010 moved from a system of warnings that were based on specific meteorological thresholds to a model that decides whether the event is likely to cause significant impact. This change was done to try to limit the number of extreme weather warnings. Less so in the United States, Europe faces the problem of multiple borders and agencies with differing rules. This places a different type of issue along the borders as some countries have different requirements for specific warnings that are passed to its citizens. Just over the border, the same weather feature will likely occur but with a different standard of warnings, and so the public may get confused or be unsure of the dangers that are coming ("Researchers Advise Warning Systems and Preparation to Mitigate the Effects of Extreme Weather Events," 2015). A more unified approach to warning systems could play a role in better preparing individuals and allowing a universal understanding of what each warning entails and how to prepare for weather events.

When it comes to nowcasting warnings, such as in tornado situations, some believe better radar technology is key. One theory is that the addition of phased-array technology on the radar will help see multiple scans instead of the one scan of Doppler radars. This will allow for multiple layers to be scanned simultaneously, and will allow meteorologists to see more rapid changes within storms. It will likely be many years before this type of technology can be put into use around the world, but it is felt that the phased array could be the next advancement to radar technology, and it is believed that this one advancement could increase tornado warning times by 18 minutes alone (Lubchenco and Hayes, 2012, pp. 1–2).

The drone is also a new and growing piece of warning technology. Drones can be used to get into hard-to-reach or dangerous locations. Drones can be used to better see where floods are spreading, to monitor landslides, or even monitor atmospheric conditions near volcanoes. Safely seeing where forest fires are and where they may be spreading is another valuable part of drone usage, helping agencies with a better understanding of where to attack fires and where they should warn and evacuate residents (Harriman and Muhlhausen, 2013). The use of drones is spreading and they have the added benefit of providing visual first-hand accounts of what is going on without putting a single life in harm's way (Figure 14.5).

Tornado warnings, often produced minutes before the tornado hits a specific location, are an area in need of advancement. One example of a way to increase tornado warning is seen with the Warning Decision Support System (WDSS). The first WDSS was installed at the NWS office at Peachtree, Georgia. This new technology includes "advanced image processing, artificial intelligence, neural networks, and other algorithms that use Doppler radar data" and has been shown to increase tornado warning lead-times as well as flash-flooding warnings and severe thunderstorm warnings by 50%

Figure 14.5 The unmanned aircraft called Sierra is part of NASA's Ames Research Center. Sierra is capable of performing remote sensing and atmospheric sampling in regions not commonly or easily accessible, such as mountaintops, open waters, or the Polar Regions. Sierra can carry over 45 kg (100 lb) and can travel up to 965 km (600 miles). *Source*: Photo courtesy of NASA.

("Severe Storms Research Center Developing New Technologies to Increase Tornado Warning Time for Georgia Citizens," n.d.). Also sirens are used in relation to tornado warnings to alert the public. The main issue with sirens is that it can never be known whether a single specific siren has complete coverage of an area. There may be locations on the edge of the siren's sound radius that are missed, leaving that area at risk. Sirens are of major importance and are a great tool to immediately alert entire towns, but the placement of more sirens, especially in tornado-prone locations, could play a big role in ensuring full coverage.

Besides these advancements, there have greater advancements in the more common methods of weather radio, television broadcasts, and alert systems, and even highway reader boards, which digitally display any weather conditions that are expected in a warned area such as winter weather warnings, tornado warnings, or evacuations (Mass, 2011). Along with these advancements is the addition of smartphone technology and usage. A great many people have a smartphone, which is often with them at all times. With this knowledge, many weather agencies have created ways to send messages and alerts via text message to the citizens in that location. Meteoalarm is one such system, using a common color scheme and symbols to indicate the severity of an extreme weather event across 32 different countries (Rogers and Tsirkunov, 2013, p. 18). Meteoalarm is available on the Internet or through applications

for the smartphone. There are detailed descriptions of the weather conditions that are forecast for a region, with links to gather more specific data from any specific weather service. The color coded system breaks weather events down into five categories: white, missing or outdated data; green, no particular awareness required; yellow, potential dangers for weather that is not unusual but needing attention; orange, dangerous weather and unusual meteorological phenomena with damages and casualties likely; and then red, very dangerous and exceptionally intense weather forecast with major damage likely and threats to life ("Meteoalarm," n.d.). The NWS also has its own online alerting system, but it does not specifically issue text messages or email alerts. Instead, private weather companies usually have the ability to disseminate weather alerts and warnings issued by the NWS to subscribers. There are dozens of companies and phone application designers that have the option to receive text alerts and warnings, all with information collected from the NWS ("Email and SMS Weather Alert Services," n.d.). Using technology that is within everyone's reach is a quick and easy way of getting information out to the public in times of need.

14.3.5 High-Resolution Models

Increasing the power and quality of high-resolution weather models is a growing demand for forecasters, especially for use during extreme weather events. Since extreme weather events

Existing weather model

New HRRR weather model

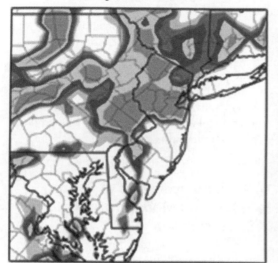

Figure 14.6 In 2014, the NWS started to implement a high-resolution weather model into its toolkit for weather forecasters and researchers. The images are of the same storm. The difference between the two models is extremely noticeable, and shows just how valuable high resolution is. As weather agencies develop more powerful computers, the better they are then equipped to run high-resolution models, greatly increasing accuracy in forecasts, even during localized thunderstorm outbreaks such as the case displayed here. *Source*: Courtesy of NOAA Photo Library. (*See color plate section for the color representation of this figure.*)

are not looked upon as a global weather feature, and instead a more localized forecast, limiting the surface area of the model allows for higher-resolution abilities. Grid spacing for some of the higher-resolution models can get closer than a 4 km (2.5-mile) resolution. These models have also a better grasp on features such as topography and changing terrains. They have better data assimilation and better initialized data. Another perk of the higher-resolution models is the idea of rapid updating, many times hourly, keeping forecasters at the forefront of nowcasting events. The US RAP model, rapid refresh, runs two versions. One is the 13 km (8-mile) resolution and the second is the 3 km (1.8-mile) resolution HRRR, high-resolution rapid refresh. RAP forecasts are generated every hour for 18-hour forecasts ("Rapid Refresh (RAP)," n.d.). The HRRR and RAP are the two most commonly used high-resolution models in the United States (Mass, 2011). The value of the HRRR is also that it covers more specific weather variables and breaks most information and modeling

down into 15-minute increments over a 15-hour period, while refreshing ever hour ("NOAA's Weather Forecasts Go Hyper-local with Next-generation Weather Model," 2014) (Figure 14.6). Some of the data modeled are radar simulations at 1 km (0.6 miles), giving a very accurate feel of future radar trends, precipitation type, precipitation rate and total, lightning forecasts, wind forecasts, visibility, and severe weather variables such as cape, helicity, and cin. With this type of high-resolution modeling, forecasters are able to focus on neighborhoods rather than regions, and can better predict where flooding, thunderstorms, tornadoes, and snow banding may occur. High-resolution models are also greatly relied upon for air-traffic forecasting.

In order to create such a powerful and important weather model, a 3D picture of the atmosphere one hour before each created forecast is put into the modeling data. Then, the addition of observations from surface stations, commercial aircraft, satellites, and weather balloons are included. This gives a very accurate picture of the atmosphere and creates a very accurate starting point for the model. Another important piece is the addition of radar data into the model. This allows the model to get a feel of where precipitation is occurring at the starting point and, with all the other information, allows for a better extrapolation over the course of the model's 15-hour run ("NOAA's Weather Forecasts Go Hyper-local with Next-generation Weather Model," 2014). The use of these types of strong high-resolution rapid refresh models allows a forecaster to be able to better warn the public based on the forecast, rather than based on detection, increasing lead-times greatly (Rogers and Tsirkunov, 2013, p. 15).

Other examples of successful high-resolution modeling include the Japan Meteorological Agency's (JMA) high-resolution model, the Local Forecast Model (LFM), which updates hourly and has a resolution of 2 km (1.2 miles), and provides a nine-hour forecast, updated every hour, for the eastern part of Japan. The Met Office runs a convective-permitting resolution model at a 1.5 km (0.9-miles) grid length, and the German weather service runs a model at a 2.8 km (1.7-mile) grid length. These additional examples of high-resolution modeling help portray mesoscale convective systems and diurnal cycles of convection. These are characteristics typical global and mid-range models miss out on (Stein, *et al.*, 2015, p. 939). Météo-France also is tapping into the high-resolution modeling scene with an hourly rapid refresh version of its AROME model, which has a 2.5 km (1.6-mile) resolution. AROME stands for Application of Research to Operations at Mesoscale and is a small-scale numerical prediction model used to improve short-range forecasts of severe events. Even stronger versions of AROME at a half kilometer are being developed and paired with the rapid refresh system for nowcasting areas of interest. This model also uses data from radar, lidar, and ground measurements. When it comes to convective weather events and other high-impact weather events, high-resolution global models have become more significant. These types of models usually have a grid spacing resolution of 1–4 km (0.6–2.5 miles), and focus on a timescale of 24 hours. In addition to convective modeling, this type of global high-resolution model also has a better toolset to forecast other local weather phenomena, such as low clouds and visibility often which is of great value to air travel ("Anticipated Advances in Numerical Weather Prediction (NWP), and the Growing Technology Gap in Weather Forecasting," n.d., pp. 8–14).

14.4 Enhancements to Radar, Satellite, and Model Output of Supercomputers

Along with forecasting and modeling technology growth, meteorology has seen and will continue to see a growth in radar and satellite technology. Aiming to improve forecasting, modeling, and warning systems for severe and extreme weather, such as thunderstorms, tornadoes, and tropical cyclones, researchers and engineers are working hard to enhance radars, satellites, and model outputs of supercomputers (Lubchenco and Hayes, 2012, pp. 1–2).

14.4.1 Radar Enhancements

One of the main issues with radar in recent years has been the lack of definition in radar images, leaving meteorologists unsure of what is being detected. This can lead to major issues when trying to distinguish whether a radar return is showing dust or rain. This was fixed when the dual-pol upgrade took place. Now, meteorologists can confidently tell the types of precipitation and/or particle size based on the images produced from the radar returns. Meteorologists also use this upgrade in tornado forecasting, when debris fields become visible near possible rotation inside thunderstorm cells. These advancements have gone a long way and have produced much better images of storm systems. The next stage in radar advancement could be the use and further development of phased-array radar, which will generate sharper images. Originally used by the US Navy to scan and detect ships and missiles, a phased-array radar would greatly decrease scan time. Current radar technology requires a scan of 360 degrees at one level of the atmosphere, then after completion of that scan the radar tilts to scan another layer. This process involves the scanning of 14 layers before returning back to the lowest layer, where it starts the process all over again. This process can take over four minutes to complete, and this leaves huge holes and dangers especially when covering severe weather outbreaks when changes can occur in seconds. The phased-array radar will be able to send out multiple beams at the same time, removing the need for the 14-layer tilting procedure, giving a complete scan in less than one minute. It is believed that this technological advancement could lead to the extension of tornado warnings by over 18 minutes. This phased-array advancement, should it be completed, will not be seen nationwide in the United States until after the year 2020 (Lubchenco and Hayes, 2012, pp. 1–2).

South Korea is also implementing radar enhancement planning. As previously stated, there is no international regulation of weather technology, leaving many countries and weather agencies alone and sometimes leading to a falling behind in technological advancements. While the United States has fully incorporated the dual-pol radars throughout the country and is now planning the next advancement, South Korea is only just integrating the dual-pol radar technology now, and plans to have all radars equipped with this technology by 2018. This upgrade is due to user demand in South Korea, and the Korea Meteorological Administration (KMA) is acknowledging its need and the benefits it can produce not only for weather aspects in regards to the increasing risks of weather disasters but also for the government as a whole. This includes departments such as the Ministry of Land in response to flood prevention, Transport and Maritime Affairs, and the Ministry of National Defense in support of military operations. Also, this will help ease the worries of all in South Korea as weather and extreme events have become a growing issue. This radar network will lead to more reliable data, and will increase the KMA's competitiveness. Unlike what we have seen throughout Europe, where there is no universal design for each radar, KMA is creating this system to have the same functions and specifications through one manufacturer, leading to a cohesive and universal radar system (Kim, Yang, and Kim, n.d., pp. 1–2).

South Korea is among many nations lagging slightly behind in radar technology and the usage of the dual-pol radar. Even the nations of Europe are still adapting to the dual-pol technological enhancement in their radar networks. For example, the Weather Radar Network Renewal Project, funded by the Met Office and the Environment Agency, is currently aiming to improve the radar system currently in place in the UK. By adding the technology of the dual-pol radar, the Met Office is increasing the capability of its radar network, improving quality control, and (of major importance to the region) increasing the accuracy of rainfall rate forecasting and nowcasting. This plan is also setting up the ability to improve further on its technology, and to extending the operational life of the network beyond the next 15 years. Some of the advancements other than the dual-pol aspect include higher-specification radomes, higher-bandwidth communications which will increase the efficiency of data generation and interpretation, and when applicable the reusing and recycling of old components to cut back on the environmental impact of this full network enhancement ("Weather Radar Network Renewal," 2013). This project is nearing completion, and is a major step toward universality of radar technology for the Met Office.

14.4.2 Satellite Enhancements

Besides radar technology, the other eye on the weather that forecasters rely on is satellite technology. Ninety percent of the data that go into a forecast are said to come from satellite data. Detailed observations that come solely from the use of satellite technology allow meteorologists to forecast at extended lengths of time, greatly increasing lead-time to potential extreme weather events. Satellites, be they geostationary or polar orbiting, both give meteorologists eyes in the sky that no other technology can provide. To help increase lead-time and extend forecasting beyond five days in regards to specific extreme weather event characteristics, the NOAA plans to "launch a new series of LEO satellites (low Earth orbit polar orbiting satellites) this decade. This is part of the Joint Polar Satellite System, with updated hardware, fitted with more sophisticated instruments" which will produce data from an improved 3D information on weather characteristics and variables such as the atmosphere's temperature, moisture, and pressure (Lubchenco and Hayes, 2012, pp. 1–2). In addition to better polar orbiting satellite technology, the United States, in 2016, has successfully launched their next series of the Geostationary Operational Environmental Satellites (GOES), with the GOES-R or the GOES-16. The GOES-R will be responsible for "continuous imagery and atmospheric measurements of Earth's Western Hemisphere, total lightning data, and space weather monitoring to provide critical atmospheric, hydrologic, oceanic, climatic, solar, and space data" ("Mission Overview," n.d.). This GOES-R enhancement will be able to provide data on weather patterns every 30 seconds. The data will create more accurate short-term forecasts and will greatly increase outlooks and watches/warnings for severe weather such as hurricanes, thunderstorms, and tornadoes. Maritime forecasts, seasonal outlooks, drought forecasts, and space weather predictions will all benefit greatly from the new technology aboard the GOES-R. The GOES-R is set up for a lifetime that extends through 2036, and is one in a series of planned advancements and launches by the NOAA, which will eventually include the GOES-S around 2017, the GOES-T around 2019, and the GOES-U around 2024 ("Mission Overview," n.d.) (Figure 14.7).

Another form of satellite enhancement is in regards to fire prediction. In the early 2000s,

NASA launched its Gravity Recovery and Climate Experiment (GRACE), where two identical spacecraft fly in tandem scanning tiny variations in earth's gravitational field, showing the movement of water over time of a given region, including soil moisture content as well as surface water and ground water amounts. The information gathered from GRACE helped lead to the first ever quantifiable data showing that dry areas do in fact coincide with increased fire danger, and it also showed that over the grasslands region an overly wet springtime results in more grass which then leads to increased fire danger during the summer. These data have given NASA scientists the ability to map out potential fire dangers in a way that shows where predicted fires may occur across the country. The success of GRACE and the introduction of new technology directed NASA to the development of GRACE-II, which is scheduled to be launched in 2017 (Skibba, 2016, p. 5). GRACE-II will have increased resolution to around 100 km (62 miles), and will be equipped with state-of-the-art technology which will increase accuracy and decrease atmospheric drag and degradation. GRACE-II is also designed to help collect data on changes in ice sheets, leading to better climate models and information on sea level rise ("Grace-II," 2016).

The European Space Agency has been very active in satellite technology as well. The Sentinel-3A was launched in 2016, to be followed by scheduled launches of Sentinel-3B in 2017, and Sentinel-3C before 2020. The Sentinel-3 is mainly an ocean mission satellite, but still with the capabilities of providing atmospheric and land information. The main objective of Sentinel-3 is to measure sea surface topography, wave height, ocean and land surface temperatures, sea and land ice topography, seawater quality, and inland water monitoring such as lakes and rivers, and aid in ocean forecasting and climate monitoring, as well as helping in fire detection and weather forecasting ("Sentinel-3," 2016). A joint mission with the United States also involved the launching of Jason-3 in 2016. Jason-3 is designed to measure

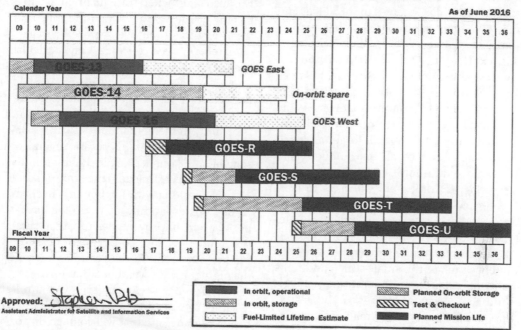

Figure 14.7 This chart shows the current geostationary satellite programs for the United States, along with the four future planned programs dating from 2016 to 2024. *Source:* Courtesy of NOAA Photo Library.

the topography of the ocean surface. These measurements provide information on ocean circulation patterns as well as climate implications and changes in sea level. The main instrument onboard Jason-3 is its radar altimeter. This device measures with extreme accuracy the sea level variations over the global ocean ("Jason-3," 2016). In 2015, the MSG-4 geostationary satellite was launched successfully and put into orbit. The European Organization for the Exploitation of Meteorological Satellites (EUMETSAT) took over the satellite shortly after its launch and now operates it for meteorological purposes. The MSG-4 will be known as METEOSAT-11, another in the series of MSG satellites. This satellite will be ready and waiting until it is needed to replace one of its aging predecessors. METEOSAT-11 will also be responsible for the transitional period prior to and during the launching and commissioning of the third generation of METEOSAT satellites, MTG, scheduled for 2019 ("MSG Overview," 2016). The MTG program will be designed to bring meteorological data accessible through satellite until the late 2030s. This generation of satellites will be based on a twin satellite concept, where there will be four imaging satellites and two sounding satellites. These sounder satellites will be the new piece that sets the EUMETSAT satellites apart from any other satellite program. The sounders will be the key to allowing the satellites to "image weather systems and analyze the atmosphere layer-by-layer" which will allow for a far greater detailed chemical composition study than ever before ("Meteosat Third Generation," n.d.).

14.4.3 Improving Model Output Data

In order to take full advantage of the new technology and new data that the radar and satellite enhancements will be able to produce and submit, computer modeling will have to be enhanced as well. It has been shown throughout this text that one of the more rapid and frequently advancing meteorological technologies concerns computer strength and speed. This pattern will have to continue alongside any other technological growth; otherwise, it will be all for nothing. The goal is to get as much data as possible, crunch all the numbers, and produce an accurate forecast and depiction of the atmosphere. Computers will have to continue to grow in order to accomplish this. Besides the speed and power, meteorologists are focusing on increasing the scaling of their models, aiming for models that will be able to "accurately simulate the small-scale conditions that catapult a routine thunderstorm or hurricane into a monster" (Lubchenco and Hayes, 2012, pp. 1–2). This type of small-scale supercomputer is in the planning stage for the United States and the hope is there that they will become more readily used and available by 2020. The director of the NOAA's Environmental Modeling Center, Bill Lapenta, believes that by 2022 the increased radar and satellite technology and capabilities will be combined with supercomputers with speeds of beyond quintillion computations per second, which then will produce computer models that will be extremely detailed and in real time (Lubchenco and Hayes, 2012, pp. 1–2). The latest crack at accomplishing Lapenta's goal came in 2015, when the United States upgraded its Global Forecasting System (GFS) forecasting model. Some of the main enhancements included increased horizontal resolution and new product fields. The resolution, possibly the greatest enhancement, increased from 27 km (17.5 miles) to 13 km (8 miles) through the first segment of the model run. The time of the first segment also extended, therefore having higher resolution and greater long-term accuracy to hour 240, rather than hour 192. The second segment of the model run also saw an increase in resolution, from 84 km (52 miles) to 35 km (22 miles). This segment was also lengthened from its original 240 hours to now 384 hours, allowing for a much further trend analysis than ever before. In regards to climatology, past formulas and calculations used data from 1982–2001. The updated formulas and calculations now use a more relevant and accurate depiction of the atmosphere and climate by extending the climatology to 2012 ("Recent Implementations," 2016). Beyond these improvements, there are several others scheduled through 2019, including the making of hourly forecasting guidance through the first 120 hours, along with further enhancements to resolution, expanding of the total forecasting period, and improvements to the accuracy of the model in the 10-day period, and even further out to the three- to four-week period during significant weather situations.

The European Center for Medium-Range Weather Forecasting (ECMWF) is seeing its own major upgrade, and that will come in the form of a $36 million contract to upgrade its Cray supercomputers. The upgrade will be complete in 2017 and will include many features that will further the claim of model dominance for the Met Office and the ECMWF. This computer will have the ability to perform more than 16,000 trillion calculations per second, and will weigh the equivalent of 11 double-decker buses. This increase in power will be 13 times greater than the current system's. This will open the doors to higher-resolution modeling, leading to more accurate and small-scale forecasting in extreme weather events. There are hopes to be able to produce a model with a resolution of just 300 m (984 ft), with the ability to forecast and model such specific details as the timing of fog over a specific airport. The supercomputer is also set to improve long-range forecasting and seasonal forecasting, and increasing the specifics in each rather than a vague analysis of precipitation and temperature ("£97m Supercomputer Makes UK World-leader in Weather and Climate Science," 2014).

14.5 Long-Range Forecasting

Nowcasting and short-range forecasting has become a dominant area of focus, especially with the growing issue of extreme weather. One other area that has seen its fair share of advancements is long-range forecasting. The World Meteorological Organization (WMO) defines a long-range forecast as a forecast done 10 days out. This can be extended to months, seasons, and years depending on the type of forecast. The process of computing long-range forecasts requires extremely large amounts of computer power, greatly limiting the amount of agencies capable of performing such tasks ("Long Range Forecasting," n.d.). The main way to increase the skill of long-range forecasting is to increase and enhance all the other aspects that go inside it. As previously stated in this chapter, enhancements in technology across the board – whether that is with water sensors, drones, radar, satellite, supercomputers, and more –will further enhance long-range modeling. All the data that are acquired through all the advanced technologies go inside models and ensembles and are then run for the period in question. With a much greater field of data for the starting point for mid- to long-range modeling, the much better the final output will be. The average of all the ensemble means can then be mapped and analyzed with much better boundary awareness, data initialization, and much less variability from run to run and within the ensembles themselves. Currently, the further out in time you go, the fewer specifics you will get on a forecast. This will be the main place of research and change in the coming years, to further the specifics and accuracies the further out in time you go.

14.6 Various Planned Advancements

Some of the main areas of focus in meteorological technology are described throughout this chapter. The following briefly describes other areas in need of research that get less attention. Some areas of focus include lidar, which is a light detection and ranging technology; increased smartphone-based weather measurement usage; cloud overlap studies; and weather modification. These areas are still of importance and should be researched further in order to make advancements in the field of meteorological technology.

14.6.1 Lidar

Lidar is similar to radar technology in the sense that it uses light from a laser to detect landscape and coastal topography. Lidar technology can be used to map and study the environment in amazing detail with hopes of leading toward a better understanding of tropical cyclones, storm surges, and global changes in climate. Lidar works by sending out laser pulses which then detect objects and accurately measure their distance to the source. Lidar imagery can be used in 3D mapping of forests and coastal areas. Sea level rise and shoreline changes can be mapped and analyzed with lidar technology. The dual wavelength Echidna lidar (DWEL), created thanks to funding from the National Science Foundation, can distinguish between laser pulse returns of wood trunks, branches, and leaves. Systems such as this have the potential to capture images of rapid coastal change from erosion and surges (Aguirre, 2015). Lidar advances can enhance surveying any changes in landscape and topography that would be a result of climate change.

14.6.2 Smartphone Measurements

Smartphones are all around us, attached to our hip, and capable of harnessing large amounts of computer power. There are plans to connect our smartphones to existing sensors for the transmitting of weather data. While there are issues to be worked out, such as inaccuracies when a smartphone is not stationary, it has been shown to be able to accurately gather and transmit the air pressure of a given location. The idea is to have smartphones equipped with pressure sensors which are then transmitted to a centralized sensor which then sends them to weather forecasting services. With the abundance of smartphones out there, the theory is that weather services will be able to discount any erroneous data received since they will be densely packed, limiting any false information. The main limitation to this idea is that the phones will have to be near to a weather station where the data will be sent. More research is needed to increase the distance from the phone to the weather station, possibly using one phone near the station as the reference point and with all other phones referencing each other. Also, other weather features are to be studied, including temperature and humidity (Zamora, Kashihara, and Yamaguchi, 2015). This type of technology will be of enormous value in places where weather observation stations are limited but the population is large, such as in India. An area like India has an infrastructure for mobile phone usage but a lack of weather instrumentation and data collection. By combining the usage of phones as weather observational instruments, a much more feasible and economical approach to expanding weather forecasting and nowcasting will become available (Senthil and Rahman, 2012, p. 186).

14.6.3 Cloud Overlap

The overlapping of clouds at different heights in the atmosphere greatly affects the amount of heat transfer and heat escape into the atmosphere. This in turn impacts the climate. A study was carried out to show the best way of simulating cloud overlap by calculating the ratio between cloud volume and cloud area at different altitudes. It was found that as the ratio became larger the overlap between the clouds at different heights became greater. It is hoped to place this theory into computer models and be used to accurately predict cloud overlap behavior in global climate models improving weather and climate forecasting (Stanley, 2016, p. 25).

14.6.4 Weather Modification

Weather modification is touched on in Chapter 10, Section 10.6.2. The advancements planned for weather modification will have a great impact on what is known and currently underway today. A major goal of weather modification is to slow the threat of climate change. The spreading of aerosols used in weather modification have been studied and shown to affect wind patterns and alter precipitation patterns (Wigington, 2014). China has the greatest investment in weather modification programs with the United States, Thailand, and India close behind. One of the main research programs in the United States is to enhance snowpack in Wyoming and Idaho (Bruintjes and Terblanche, 2015, pp. 4–5). The effects of cloud seeding have been modeled by the NCAR WRF model, helping with the design and evaluation of cloud seeding efforts (Bruintjes and Terblanche, 2015, p. 6). Research is being conducted in Japan, China, Australia, the United States, and elsewhere to design new ratios of aerosol usage, the effects it has on weather, and with new tools such as dual-pol radars, weather models, and satellites, researchers are able to now better quantify the role they are playing in modification of weather systems. As better modeling, radar, and satellite technology continue to come on line, better results and methodology of weather modification will occur (Figure 14.8).

While there is abundant proof of weather modification and cloud seeding effecting precipitation, the amount of precipitation is often negligible. Most of the time, cloud seeding helps create a more stratiform type cloud, one typical of light drizzle or snowfall. These types of precipitation producing clouds often have the precipitation that falls and evaporates before it reaches the ground. In cases like this, it is more common that cloud seeding creates clouds, and very light precipitation. For the heavier precipitation, a stratocumulus type cloud is needed. In order to create this type of cloud, the perfect setup, makeup of chemicals, and perfect timing and placement has to occur. This can and has

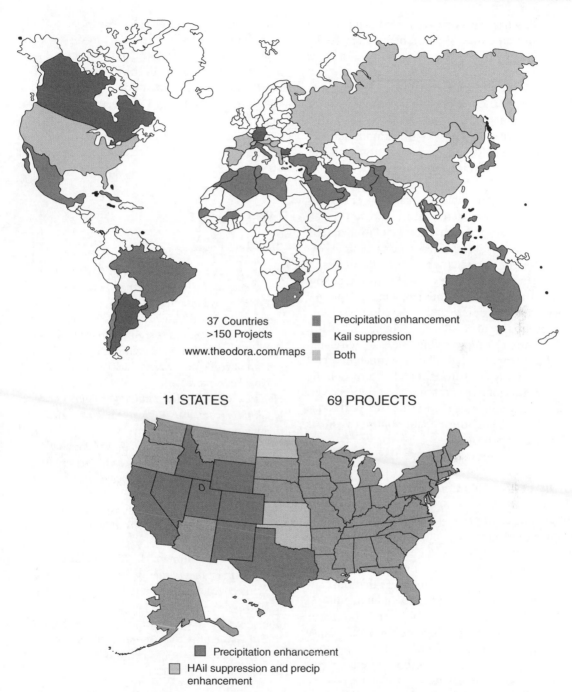

37 Countries
>150 Projects

www.theodora.com/maps

- Precipitation enhancement
- Kail suppression
- Both

11 STATES   69 PROJECTS

- Precipitation enhancement
- HAil suppression and precip enhancement

Figure 14.8 The top image shows the current weather modification programs worldwide, including those that participate in precipitation enhancement, hail suppression, or both. In total, 37 counties and over 150 projects are currently ongoing. The bottom image reflects the current weather modification programs operating in the United States alone, including 11 total states and 69 total projects. *Source*: Reproduced with permission of UCAR.

been shown to have occurred, but the likelihood of large-scale precipitation being created by artificial cloud seeding is remote, and it is unlikely that it would have much global impact even if it did (Goodell, 2010, pp. 172–73).

There are other views emerging about weather modification. One opposite weather-related opinion is that weather modification is actually causing climate change. Some scientists feel not only that the chemicals used in weather modification are extremely toxic and detrimental to the health of humans but also that these chemicals and aerosols are damaging the ozone layer. They also feel it is shrinking the atmosphere, destroying certain species, and causing excessive drought in areas while creating excessive rainfall in others. Those against it also state how data show a global dimming or a loss of blue sky during the day due to an increase in particulate matter in the atmosphere (Wigington, 2014). The dimming or loss of blue sky can be related to the change in light being refracted off of growing particles throughout the atmosphere, in turn altering what the human eye perceives. While science can be shown to prove both sides of the argument, the debate continues and is expected to continue.

Chapter Summary

Technological advancements in weather have been witnessed throughout human history. Each advancement represents a change that was once inconceivable. Today is no different, with advancements and plans underway that will continue to change the science and understanding of meteorology, our climate, and our atmosphere. Some of these changes include the tools we use to gather information such as drones, sensors, radar, satellites, and supercomputers. These tools are used to advance meteorological reaches and capabilities with data collection and weather forecasting. They also aid with nowcasting and improvements in real-time weather data. Few would have ever dreamed of some of these advancements, and the information they have

the potential to deliver is almost inconceivable. The field of meteorology is a forever-evolving science, with new areas of concern popping up and technology to meet those concerns developed and expanded upon. With extreme weather and nowcasting becoming part of our everyday life, and new ideas and research exploding all around us, meteorology today is like nothing we would have ever dreamed of just 10 years ago.

References

Book References

Acton, A. (2013) *Issues in Global Environment: Climate and Climate Change*, Scholarly Editions, Atlanta, GA.

Goodell, J. (2010) *How to Cool the Planet: Geoengineering and the Audacious Quest to Fix Earth's Climate*, Houghton Mifflin Harcourt, Boston, Massachusetts and New York, New York.

Henson, R. (2014) *The Thinking Person's Guide to Climate Change*, American Meteorological Society, Boston.

Rogers, D. and Tsirkunov, V. (2013) *Weather and Climate Resilience: Effective Preparedness through National Meteorological and Hydrological Services*, The International Bank for Reconstruction and Development/The World Bank, Washington DC.

Senthil, K. and Rahman, H. (2012) *Mobile Computing Techniques in Emerging Markets: Systems*, Applications and Services, IGI Global, Hersey, PA.

Sobel, A. (2014) *Storm Surge: Hurricane Sandy, Our Changing Climate, and Extreme Weather of the Past and Future*, HarperCollins Publishers, New York.

Journal/Report References

"A Decade of Progress in Projections and Modeling." (2015) *Hurricane Katrina: 10 Years Later*. NOAA.

Antonescu, B., Schultz, D., Lomas, F., and Kuhne, T. (2016) Tornadoes in Europe: Synthesis of the observational datasets. *Monthly Weather Review, AMS.* 7 January 2016. Abstract.

Baxter, R. and Bush, D. (2014) Use of Small Unmanned Aerial Vehicles for Air Quality and Meteorological Measurements. *2014 National Ambient Air Monitoring Conference,* T&B Systems, Inc. Valencia, CA.

Bruintjes, R. and Terblanche, D. (2015) Report of the Expert Team on Weather Modification Meeting. *WMO & ETMW,* Phitsanulok, Thailand, 17–19 March 2015, pp. 4–6.

Cheng, V., Arhonditsis, G., Sills, D., *et al.* (2016) Predicting the Climatology of Tornado Occurrences in North America with a Bayesian Hierarchical Modeling Framework. *Journal of Climate,* 29 (5), Abstract.

Do Hoai, N., Udo, K., and Mano, A. (2011). Downscaling Global Weather Forecast Outputs Using ANN for Flood Prediction. *Journal of Applied Mathematics: 2011,* Article ID 246286, pp. 1–14.

Francesch-Huidobro, M., Dabrowski, M., Tai, Y., *et al.* (2015) Governance challenges of flood-prone delta cities: Integrating flood risk management and climate change in spatial planning. *Progress in Planning,* http://www.sciencedirect.com/science/article/pii/S0305900615000628. 18 March 2016.

Ginis, I. and Bender, M. (2013) Improving the Operational Tropical Cyclone Models at NOAA/NCEP and Navy/FNMOC. *Final Report for the OAR Office of Weather and Air Quality (OWAQ),* University of Rhode Island, August 2011–July 2013.

Hall, J., Evans, E., Penning-Rowsell, E., *et al.* (2003) Quantified scenarios analysis of drivers and impacts of changing flood risk in England and Wales: 2030–2100. *Global Environment Change Part B: Environmental Hazards,* 5 (3–4), Abstract.

Harriman, L. and Muhlhausen, J. (2013) A New Eye in the Sky: Eco Drones. *UNEP Global Environment Alert (GEAS).*

Kim, J., Yang, J., and Kim, S. (n.d.) Enhancement of Korean Weather Radar Capability by Introducing a Dual-Pol Radar Network. *Radar Operation,* Weather Radar Center, Korea Meteorological Administration, Seoul.

Mass, C. (2011) Nowcasting: The Next Revolution in Weather Prediction. *Department of Atmospheric Sciences, University of Washington,* Bulletin of the American Meteorological Society, July 2011.

Neumann, B., Vafeidis, A., Zimmermann, J., and Nicholls, R. (2015). Future Coastal Population Growth and Exposure to Sea-Level Rise and Coastal Flooding: A Global Assessment. *PLOS ONE,* 10 (3), e0118571.

"Precipitation Nowcasting." (n.d.) *National Center for Atmospheric Research,* NCAR.

Skibba, R. (2016) Assessing US Fire Risks Using Soil Moisture Satellite Data. *EOS,* 97 (3), p. 5.

Stanley, S. (2016) Cloud Overlap Observations Put Simulations to the Test. *EOS,* 97 (3), p. 25.

Stein, T., Hogan, R., Clark, P., *et al.* (2015) The Dymecs Project: A Statistical Approach for the Evaluation of Convective Storms in High-Resolution NWP Models. *BAMS,* 96 (6), p. 939.

Summer, T. (2015) The Future of Forecasting: Technology Promises faster Weather Predictions on a Smaller Scale. *Science News, Society for Science & the Public,* 187 (9), pp. 20–23.

Wang, J., Young, K., Hock, T., *et al.* (2015) A Long-Term, High Quality, High-Vertical-Resolution GPS Dropsonde Dataset for Hurricane and Other Studies. *BAMS,* 96 (6), pp. 970–971.

Zamora, J., Kashihara, S., and Yamaguchi, S. (2015). Calibration of Smartphone-Based Weather Measurements Using Pairwise Gossip. *The Scientific World Journal,* Article ID 494687, Abstract.

Website References

"£97m Supercomputer Makes UK World-Leader in Weather and Climate Science." (2014) *Met Office,* http://www.metoffice.gov.uk/news/releases/archive/2014/new-hpc, accessed 3 April 2016.

"Advancements in Detection, Forecasting and Safety Result from 2004 Indian Ocean Tsunami." (2014) *Weather-Ready Nation*, NWS & NOAA, http://www.nws.noaa.gov/com/weatherreadynation/news/141203_tsunami_story.html#.Vqjxd_krIkV, accessed 19 March 2016.

Aguirre, E.L. (2015) *Lidar Will Be Used to Study Forests, Wetlands, Sea Coasts*, UMass Lowell, https://www.uml.edu/News/stories/2015/Lidar-technology.aspx, accessed 3 April 2016.

"Anticipated Advances in Numerical Weather Prediction (NWP), and the Growing Technology Gap in Weather Forecasting." (n.d.) *WMO, with contributions from the CBS/DPFS Chairperson*, https://www.wmo.int/pages/prog/www/swfdp/Meetings/documents/Advances_NWP.pdf, accessed 1 March 2016, pp. 8–14.

"Coastal Flood Maps." (2016) *Floodsmart.gov*, National Flood Insurance Program, https://www.floodsmart.gov/floodsmart/pages/coastal_flooding/coastal_flood_maps.jsp, accessed 30 March 2016.

"Email and SMS Weather Alert Services." (n.d.) *NWS*, NWS & NOAA, http://www.weather.gov/subscribe, accessed 22 March 2016.

Feibel, C. (2014) *NOAA Brings 'Hurricane Hunter' Airplane to Houston*, Houston Public Media, http://www.houstonpublicmedia.org/articles/news/2014/05/19/51111/noaa-brings-hurricane-hunter-airplane-to-houston, accessed 15 March 2016.

"Grace-II." (2016) *Gravity Recovery and Climate Experiment II – Earth Science and Applications from Space*, Satellite Observations to Benefit Science and Society: Recommended Missions for the Next Decade, http://www.nap.edu/read/11952/chapter/12, accessed 22 March 2016.

Hannam, P. (2015) *Australia Faces Rising Perils From Climate Change, Earthquakes: Munich Report*, The Sydney Morning Herald: Environment, 9 November 2015, http://www.smh.com.au/environment/climate-change/australia-faces-rising-perils-from-climate-change-earthquakes-munich-re-report-20151109-gku2p0.html, accessed 15 March 2016.

"Hurricane Hunters." (2015) *Hurricanes: Science and Society*, http://www.hurricanescience.org/science/observation/aircraftrecon/hurricanehunters, accessed 15 March 2016.

"Hurricane Scientists Bring a New Wave of Technology to Improve Forecasts." (2016) *AOML*, NOAA & Atlantic Oceanographic & Meteorological Laboratory, http://www.aoml.noaa.gov/keynotes/keynotes_0814_new_hurricane_research.html, accessed 15 March 2016.

"HWRF: About Us." (2014) *HWRF: The Hurricane Weather Research and Forecasting System*, NOAA & Atlantic Oceanographic & Meteorological Laboratory, http://hwrf.aoml.noaa.gov/about, accessed 16 March 2016.

"Jason-3." (2016) *Ocean Surface Topography from Space*, NASA Jet Propulsion Laboratory California Institute of Technology, http://sealevel.jpl.nasa.gov/missions/jason3, accessed 22 March 2016.

"Long Range Forecasting." (n.d.) *World Meteorological Organization*, https://www.wmo.int/pages/themes/climate/long_range_forecasting.php, accessed 2 April 2016.

Lubchenco, J. and Hayes, J. (2012) *New Technology Allows Better Extreme Weather Forecasts*, Scientific American, http://www.scientificamerican.com/article/a-better-eye-on-the-storm, accessed 22 March 2016. pp. 1–2.

"Meteoalarm." (n.d.) *Meteoalarm – Alerting Europe for Extreme Weather, Meteoalarm.eu*, http://www.meteoalarm.eu/about.php?lang=en_UK, accessed 22 March 2016.

"Meteosat Third Generation." (n.d.) *EUMETSAT*, MTG, http://www.eumetsat.int/website/home/Satellites/FutureSatellites/MeteosatThirdGeneration/index.html, accessed 27 March 2016.

"Mission Overview." (n.d.) *GOES-R*, NOAA & NASA, http://www.goes-r.gov/mission/mission.html, accessed 24 March 2016.

"MSG Overview." (2016) *European Space Agency*, MSG, http://m.esa.int/Our_Activities/Observing_the_Earth/Meteosat_Second_Generation/MSG_overview2, accessed 22 March 2016.

"NASA's Global Hawk Drone Aircraft Flies over Frank on the GRIP Hurricane Mission." (2010) *GRIP Hurricane Mission*, NASA, http://www.nasa.gov/mission_pages/hurricanes/missions/grip/news/frank-flyover.html, accessed 15 March 2016.

"New NOAA Lake Pontchartrain Sensors to Provide Better Evacuation Planning, Storm Surge Data." (2015) *NOAA News*, NOAA, http://www.noaanews.noaa.gov/stories2015/092315-new-noaa-lake-pontchartrain-sensors-to-provide-better-evacuation-planning-storm-surge-data.html, accessed 19 March 2016.

"New Report Finds Human-Caused Climate Change Increased the Severity of Many Extreme Events in 2014." (2015) *NOAA News*, NOAA, http://www.noaanews.noaa.gov/stories2015/110515-new-report-human-caused-climate-change-increased-the-severity-of-many-extreme-events-in-2014.html, accessed 15 March 2016.

"NOAA's Weather Forecasts Go Hyper-local with Next-generation Weather Model." (2014) *NOAA News*, NOAA, http://www.noaanews.noaa.gov/stories2014/20140930_hrrr.html, accessed 23 March 2016.

"Nowcasting." (n.d.) *Public Weather Service (PWS)*, WMO, https://www.wmo.int/pages/prog/amp/pwsp/Nowcasting.htm, accessed 21 March 2016.

"Rapid Refresh (RAP)." (n.d.) *National Centers for Environmental Information*, NOAA, https://www.ncdc.noaa.gov/data-access/model-data/model-datasets/rapid-refresh-rap, accessed 23 March 2016.

"Recent Implementations." (2016) *Environmental Modeling Center*, NWS & NOAA, http://www.emc.ncep.noaa.gov/GFS/impl.php, accessed 3 April 2016.

"Researchers Advise Warning Systems and Preparation to Mitigate the Effects of Extreme Weather Events." (2015) *Phys Org*, http://phys.org/news/2015-12-mitigate-effects-extreme-weather-events.html, accessed 22 March 2016.

"Sentinel-3." (2016) *ESA EO Missions - Earth Online*, https://earth.esa.int/web/guest/missions/esa-future-missions/sentinel-3, accessed 22 March 2016.

"Severe Storms Research Center Developing New Technologies to Increase Tornado Warning Time for Georgia Citizens." (n.d.) *Georgia Tech Research Institute*, http://www.gtri.gatech.edu/casestudy/severe-storms-research-center, accessed 22 March 2016.

"Storm Surge and Coastal Inundation." (n.d.) *NOAA*, http://www.stormsurge.noaa.gov/models_obs_monitoring.html, accessed 19 March 2016.

"Thunderstorm Auto Nowcasting: Overview." (2016) *Thunderstorm Auto Nowcasting: RAL*, NCAR, https://www.ral.ucar.edu/projects/nowcast, accessed 21 March 2016.

"USGS Surge, Wave, and Tide Hydrodynamics (SWATH) Network." (2016) *USGS Flood Information*, USGS, http://water.usgs.gov/floods/swath, accessed March 19, 2016.

"WCRP: Home." (2016) *WCRP World Climate Research Programme*, http://wcrp-climate.org, accessed 16 March 2016.

"Weather Radar Network Renewal." (2013) *Met Office*, http://www.metoffice.gov.uk/water/radarimprovements, accessed 24 March 2016.

Wigington, D. (2014) Global Weather Modification Assault Causing Climate Chaos and Environmental Catastrophe, Geoengineering Watch, http://www.geoengineeringwatch.org/global-weather-modification-assault-causing-climate-chaos-and-environmental-catastrophe-2/, accessed 18 March 2016.

Figure References – In Order of Appearance

"NOAA Photo Library." (n.d.) *NOAA*, http://www.photolib.noaa.gov/htmls/wea01803.htm. 18 August 2016.

"NOAA Hurricane Hunters." (2016) *Aircraft Operations*. NOAA, http://www.omao.noaa.gov/learn/aircraft-operations/about/hurricane-hunters, accessed 22 July 2016.

"Recognizing Sandy 2012." (2014) *Ocean and Atmospheric Research*. NOAA, http://research.noaa.gov/News/NewsArchive/LatestNews/TabId/684/ArtMID/1768/ArticleID/10867/Recognizing-Sandy-2012.aspx, accessed 20 July 2016.

"USGS Surge, Wave, and Tide Hydrodynamics (SWATH) Network." (2016) *USGS Flood Information*. USGS, http://water.usgs.gov/floods/swath, accessed 19 March 2016.

"New NOAA Lake Pontchartrain Sensors to Provide Better Evacuation Planning, Storm Surge Data." (2015) *NOAA News*. NOAA, http://www.noaanews.noaa.gov/stories2015/092315-new-noaa-lake-pontchartrain-sensors-to-provide-better-evacuation-planning-storm-surge-data.html, accessed 19 March 2016.

"NASA Airborne Science Program." (2015.) *NASA*. Page Editor Fladeland, M., https://airbornescience.nasa.gov/aircraft/SIERRA, accessed 18 July 2016.

"NOAA's Weather Forecasts Go Hyper-local with Next-generation Weather Model." (2014) *NOAA News*. NOAA, http://www.noaanews.noaa.gov/stories2014/20140930_hrrr.html, accessed 23 March 2016.

"Mission Overview." (n.d.) *GOES-R*. NOAA & NASA, http://www.goes-r.gov/mission/mission.html, accessed 24 March 2016.

"Weather Modification – Multimedia Gallery." (2015) *University Corporation for Atmospheric Research*. NCAR & UCAR, https://www2.ucar.edu/news/weather-modification-multimedia-gallery, accessed 16 September 2015.

15

Global Cooperation

Kevin Anthony Teague

It is clear that the science of meteorology has advanced and will always be advancing to meet the present-day needs of humankind. Throughout history it has been shown that individualized approaches toward advancement have been the dominant form of advancing this field, whether through competition between nations, war, economic reasons, natural disasters, or just individual priorities. This chapter explains how this approach has begun to change through global cooperation. This new direction will have a much more powerful and far-reaching influence on humankind than any sort of individualized way of thinking.

15.1 Overall Global Forecasting

It is extremely obvious that there are extreme weather events occurring each and every year: record-breaking warmth or cold, record rainfall or snowstorm totals, record tropical cyclone strengths and coastal flooding, record tornado outbreaks. This list keeps growing and the occurrences are never ending. A clear constant in extreme weather is that it is a global issue. Events do not just happen in the United States, or in Europe. Instead, extreme weather events occur in every nook and cranny of earth. Some of the more devastating occurrences actually happen where there are large populations and low economies. This is usually due to a lower quality of living conditions, lower governmental outreach, and much lower standards of meteorological forecasting and warning. It isn't that extreme weather events gravitate to these locations; it is instead that any event which occurs becomes an extreme event due to the standards put in place for these locations. Beyond that, it has been well documented throughout this text of the past and more recent extreme weather events that have occurred across the world. In response to these events and other variables, meteorology has been exploding with advancements. The United States has implemented new sensor technologies to monitor storm surge events, the South Koreans have updated their radar technology, and the Europeans have made huge advancements in their supercomputer capabilities and their weather modeling that dominates the forecasting world. This is just a very small sample of the advancements that have occurred. The problem with all these advancements is that they are within their own borders or agencies. Instead, attention has more recently began to turn toward a more cooperative approach, stretching these and other advancements across the globe, with countries working together instead of competing against each other.

The Evolution of Meteorology: A Look into the Past, Present, and Future of Weather Forecasting, First Edition.
Kevin Anthony Teague and Nicole Gallicchio.
© 2017 John Wiley & Sons Ltd. Published 2017 by John Wiley & Sons Ltd.

With extreme weather events affecting earth as a whole, scientists across the globe agree that in order to take the world closer to a level playing field in terms of forecasting and data collection a lot of attention needs to be placed on data sharing. As stated in Chapter 9, Section 9.4.3, many nations in Southeast Asia have agreed to data sharing in regards to radar data to help observe and forecast weather systems across a greater distance to keep from being caught off guard. Sharing data across borders is a crucial building block to global involvement, especially in the less economically developed regions across the world. Financial agreements and technical capabilities become issues governments have to work through in order to protect lives and livelihoods from extreme weather disasters. Data sharing can involve radar sharing, as done in Australia with its neighboring nations, or by satellite sharing, NWP (numerical weather prediction) data sharing, model sharing, forecast sharing, etc. In regards to global data, some weather agencies produce global forecasting models. This information can be shared across borders, especially in areas that have a very limited capability to produce accurate and dependable forecasting models for their own regions. These global data can be shared and used by localized weather departments to improve their regionalized forecast. In order to share these global data, telecoms would have to be in place across the globe, especially in less populated areas, such as the mountainous regions of the Balkan Peninsula or Central Asia. Manual observations are being replaced around the world by a more computerized and reliable automatic observation. This type of observation is easily transmittable when it comes to data sharing. Improving observation standards and telecom units for the transporting of data across the globe will help global models become more accurate. More weather observation points across the world will improve weather models in setting up the most accurate depiction of the current state of the atmosphere. It will improve all the boundary conditions, draw a more complete sampling of the earth's atmosphere, and will enhance the local accuracy of all forecasts. The other factor necessary for improved globalized forecasting is the training of forecasters themselves. In order for any data sharing of any kind to be worthwhile, the forecaster receiving that information and conveying it as a forecast for the public needs to have proper training, and an education in and understanding of the task of creating and transmitting a weather forecast (World Bank, 2008, pp. 35, 38–46, 55–56).

An example of an international, yet regionalized, cooperation and sharing of data and resources would be the e-ASIA Joint Research Program. This is a collaboration of 16 member organizations from 12 countries, coming together to promote innovation and to resolve shared weather challenges in the East Asian region. The J-RAPID Program of the Japan Science and Technology Agency (JST) is a research and investigative collaboration of Japan and other foreign researchers in response to natural or anthropogenic disasters, such as earthquakes and tropical cyclones. Programs such as these two examples show a great response to the need of international collaboration and sharing of data, ideas, and research in order to rehabilitate regions suffering from extreme weather events and to help improve the reduction of potential future disaster risk ("International Workshop on Disaster Risk Reduction and Management," 2015).

Cooperation more on the global scale would be programs through the World Weather Research Program (WWRP) and the World Climate Research Program (WCRP). These two programs helped create the Subseasonal to Seasonal Prediction Initiative, which aims to "improve forecast skill and enhance knowledge of processes on the subseasonal to seasonal timescale with a focus on the risk of extreme weather, including tropical cyclones, droughts, floods, heat waves, and the waxing and waning of monsoon precipitation" ("The World Weather Research Programme: A 10-year Vision," 2015). Having a better grasp on subseasonal forecasting will greatly help forecast skill and emphasis on high-impact weather events that often disrupt and create major issues with agriculture, safety, and water, and reduce disaster risk and health implications. Another project associated with the WWRP is the High Impact Weather (HIWeather) project. The goal of this project is to "promote cooperative international research to achieve a dramatic increase in resilience to high-impact weather worldwide by improving forecasts for timescales of minutes to two weeks and enhancing their communication and utility in social, economic, and environmental applications" ("The World Weather Research Programme: A 10-year Vision," 2015). The HIWeather project tries to emphasize that the impact of a weather event should be understood not just in terms of the severity of the extreme weather event itself but also in terms of the vulnerability of those exposed to the extreme weather event. HIWeather states that "in order to increase resilience, research is required to improve the monitoring and prediction of weather and related hazards, but also to better understand the human impacts and effectively communicate information to those most vulnerable" ("The World Weather Research Programme: A 10-year Vision," 2015). While better sharing of data and better global involvement in extreme weather forecasting and prediction are important and popular areas of global cooperation, it is also important to cooperate on studies and data from previous extreme weather events. There is a need to combine resources across the globe and to study the impacts of previous storm and weather events. This will give a better description and comprehension of the full issue. Whether this will better show the economic toll of extreme weather or the casualties related to extreme weather, or help show the at-risk locations for extreme weather, the studying of past weather in all regions is needed (World Bank, 2008, p. 73).

The World Meteorological Organization (WMO) is one of the greatest global contributors to international cooperation. The WMO is a global network that takes data from satellites, aircraft, ocean buoys, ships, radar, surface stations, weather balloons, and more from all around the world. In order for the WMO to manage and analyze all this incoming data, a process coordinated by the World Weather Watch (WWW) combines everything to get a complete picture of the atmosphere, oceans, and land surface. This information is used to initialize and constrain NWP models (Rogers and Tsirkunov, 2013, pp. 31–35).

15.2   Global Networking in Meteorology

While global forecasting and climate change are major areas of focus for global cooperation in meteorology, the science has seen an all-round coming together which has introduced new ideas, theories, technologies, and knowledge never seen before. This trend of a general cooperation in the field of meteorology is set to continue to grow in the foreseeable future. There are world congress expos and international conferences scheduled across the world with topics covering biodiversity, pollution, coastal zones, earth science, climate change, recycling, green energy, geology, oceanography, remote sensing, geophysics, and geosciences. These expos and conferences are scheduled to take place in Madrid, Dubai, Bangkok, Berlin, Atlanta, Brisbane, New Orleans, Toronto, and Valencia just to name a few. This shows that these topics of concern and value are important to people of every nation and every part of the world ("Earth & Environmental Sciences Journals," 2016). The United Nations has a very powerful and far-reaching platform with goals set each year ranging in topics from AIDS/HIV help and research to ending poverty and hunger, providing all with a good education and help and awareness in many more important and valuable areas of need. Sustainable development goals are set annually by the United Nations in order to aim for a transformation of specific policies within the world. For many years now, these goals have included climate, sustainable energy and clean energy, and life on land and in water as major platforms ("Sustainable Development Goals," 2016).

Nations are coming together to tackle more and more of the pressing needs regarding meteorology and the effects it has across the world. There are not many topics that most nations can agree on that affect everyone and can be understood by everyone. The United Nations, the WMO, and other worldwide agencies and departments confront these issues and make them become one issue for the world.

Throughout the world, agreements between agencies are taking place to allow poorer countries to share research, technologies, and resources with their neighbors. Australia, East Timor, and 14 Pacific Island countries have drawn up the Pacific-Australia Climate Change Science and Adaptation Planning (PACCSAP) program, to bring the whole region together to deal with the issue of future climate risks. These countries see the value in exchanging group knowledge rather than one nation succeeding and leaving other lesser-developed nations struggling on their own ("Collaborative Projects," n.d.). The PACCSAP was established by the Australian government's International Climate Change Adaption Initiative (ICCAI). Australia being the main nation within this program made sure to meet the priorities of the vulnerable partner countries and did so by spending hundreds of millions of dollars ("About Us," n.d.). In Europe, the EUMETNET EIG, the Network of European Meteorological Services Economic Interest Group, is a group of 31 European national meteorology agencies and departments which strives "to help its Members to develop and share their individual and joint capabilities through cooperation programs that enable enhanced networking, interoperability, optimization and integration within Europe; and also to enable collective representation with European bodies in order that these capabilities can be exploited effectively" ("EUMETNET and EU: About Us," n.d.).

Throughout the world, agreements through organizations such as the WMO, or regional agreements such as the EUMETNET EIG, are made to help the greater good and the public rather than the individual. As stated numerous times, weather happens everywhere, and in order to get an as-perfect-as-can-be depiction of the state of the atmosphere across the globe, worldwide data collection and sharing has to take place. The SIMDAT project is a virtual global information and systems center (VGISC) developed by the WMO. The objective of SIMDAT is to help in the development of "a virtual and consistent view of all meteorological data distributed in the real-time and archived databases of the partners and to provide a secure, reliable and efficient mechanism to collect, exchange and share these distributed data, in order to support research and operational activities of the meteorological community" ("Pilot Projects and Related Initiatives," n.d.). The state-of-the-art technology behind this project and the VGISC infrastructure will allow the collection of masses of information that will become accessible to a diverse user group. Interested users range from Australia, China, Japan, South Korea, and Russia, with data sets and meteorological information being added to the portal by Asia, Australia, Europe, and the United States (Raoult, Thole, and Gartel-Zafiris, 2016). This example of global sharing of data represents perfectly the value of such collaboration between nations and continents.

Global cooperation isn't only set on data sharing for forecasting purposes. It is also extremely valuable and necessary for global research. The international organization World Climate Research Program (WCRP) has identified five grand challenges in climate and meteorology: clouds and atmospheric circulation, regional sea level, climate extremes, water availability, and rapid cryosphere changes. These challenges help produce the need for periodic climate assessments, such as the Lessons Learnt for Climate Change Research meeting in Switzerland in 2014. At this meeting, 75 researchers came together to discuss these challenges but also to evaluate climate science as a whole, create new directions and plans for the WCRP, and discuss the future needs of further research and assessments. This meeting also led to the discussion of four major expansions to the original challenges presented by the WCRP. These areas focused on the need for more research and knowledge in ocean heating and circulation particularly in the deep ocean regions. Other topics of new research and development include: annual to decadal timescales in regards to water availability, short-lived climate forces such as the influences of aerosols, and the need to incorporate biogeochemical cycles into analyses and models. Almost all the researchers at the Lessons Learnt meeting explained the dire need to have a better and more systematic source of and access to data. Recognizing the extremely positive impact of meteorological reanalysis across and beyond atmospheric research and modeling, we anticipate movement by the major modeling centers toward broader earth system reanalysis. The gathering of information across various science fields, geographical locations, and ecological communities, and then having it processed into a uniform format for all, is of utmost importance in today's world. It is recognized that there are "substantial technical challenges arising from variable spatial resolutions and temporal extents," but it is argued that "such a planetary diagnosis effort represents a long-avoided task whose implementation would reverberate strongly through science and data communities" (Brasseur and Carlson, 2015). Whether the world comes together to help with research or modeling, forecasting or assessment, global cooperation throughout the field of meteorology has been proven to be effective and the most efficient, cost-effective, reliable, and far-reaching way of getting meteorological information to as many weather or climate agencies or organizations as possible.

15.3 Disaster Management

Disaster management has become a worldwide trend, with larger nations assisting less-developed nations, and fellow power nations helping each other. This help comes as aid, but also in the form of research and the spreading of knowledge and policy. There are sometimes issues regarding disaster management such as when neighboring nations have different ways of giving out warnings of storms and extreme weather events, leaving the public confused and making inappropriate preparations for the impending storm. Global cooperation can limit the adverse impact of such issues, making policy ideas, research, risk assessment, and aid more universal and uniform.

Natural hazards are defined by the WMO as severe and extreme weather and climate events that occur worldwide. A natural disaster is when a natural hazard destroys lives, livelihoods, and property. The WMO explains that the best way to prepare the public is to provide an accurate forecast and warning system that is easily understood, and further by educating the public on preparedness and management before and during events. The WMO concludes that every one dollar spent on preparedness will save seven dollars of economic loss caused by a weather-related disaster. By 2019, the WMO hopes to reduce the average 10-year fatality rate that was seen from 1994 through 2013 by 50% for weather- and climate-related disasters. Whether these disasters are short-lived events, such as those of tornadoes or flash floods, or others that are of a longer duration, such as droughts, each has its own unique issues ("Natural Hazards," n.d.). Certain regions of the world are more prone to monsoonal patterns, tropical cyclone impacts, fire, wind damage, lightning, blizzards, drought, hail, avalanches, mudslides, etc. Droughts are caused by the deficiency of rainfall or snowfall in mountainous regions responsible for a region's summertime weather, such as in California. Droughts develop slowly, sometimes over the course of seasons or years, and result in a lack of water, crop destruction, animal and human death, and malnutrition. Sometimes drought conditions cause flooding. This sometimes occurs when there is sudden heavy rainfall on a dry surface where water cannot penetrate, leading to a flash flood. Flash flooding can and does take place in any part of the world. Flash floods can be due to storms, weather patterns, monsoonal rains, snowmelt, dam or levee breaks, ice jams, high tides, or poor drainage systems. The WMO states that over 1.5 billion people were affected by floods in the 1990s, nearly 25% of the world's population. The WMO has set up its very own program in response to the devastation caused by tropical cyclones. Air pollution can be caused by industries, vehicles, and various human activities, but become a weather hazard when certain weather characteristics take place, which leads to an inversion in the lower levels of the atmosphere, blocking any escape of the pollutants into the atmosphere and instead trapping them at the surface. The WMO collects and observes pollutants in the atmosphere via its Global Atmospheric Watch (GAW) program. Mudslides and avalanches are localized events and are usually unexpected, although there are signs pointing to the potential of their occurrence. Landslides can move at over 50 km/h (31 mph) and avalanches at over 150 km/h (93 mph). Landslides tend to occur when heavy rain or rapid snowmelt takes large amounts of earth with it as it flows down a hill or mountain. Avalanches occur when shifts in snow surface due to vibration or weight cause masses of snow and ice to break and fall down a mountain taking everything in its path along with it. Dust storms occur in the United States, Africa, Australia, the Middle East, and China, threatening lives, health, and transportation as dust and sand get caught up by the wind to be transported over entire regions. Heatwaves, arctic outbreaks, tornadoes, severe storms, lightning, and blizzards have been researched in great detail by scientists around the world. Each type of these different weather phenomena comes with its own issues, its own level of predictability, and its own program of research, warning, forecasting, and management ("Natural Hazards," n.d.). See Appendix Figures C–F for more statistics on loss events worldwide from 1980 to 2015.

J-RAPID's collaboration between researchers around the world with a focus on East and Southeast Asia has produced some very important work and has contributed greatly to the field of disaster management. J-RAPID provides initial response to disasters, oftentimes before other academic organizations or even national governments. With full international collaboration, J-RAPID performed admirably in response to the 2015 Nepal earthquake, in 2014 in response to Typhoon Yolanda, in 2012 in response to the Thai flood disaster, and in 2011 regarding the catastrophic 2011 Japan earthquake. Some research projects that J-RAPID is collaborating on with the United States reflect the value and importance in humanitarian logistics, recovery projects using underwater robots for tsunami disasters, building infrastructure to withstand tsunami waters and debris, nuclear radiation during severe storm accidents, evaluation of bridge infrastructure in regard to earthquake threats, aerial robots for exploration and mapping, and social networking services during post-catastrophe response. The United Kingdom has worked with J-RAPID in regards to the surveying of the washout effect of radioactive materials by rainfall on land surfaces. Indonesia has also worked on this project to help with evacuation research for responses to unexpected tsunamis. This global relationship is a perfect example of researchers coming together to help fight regional issues and extreme conditions ("J-RAPID," 2012).

While J-RAPID focuses mainly on a specific region in the world, other organizations approach disaster management globally. Such organizations include Disaster Preparedness and Emergency Response Association (DERA), the World Health Organization (WHO), and the Red Cross and Red Crescent networks. The Red Cross is probably one of the more recognizable organizations when it comes to combating, and relief efforts in response to, humanitarian crisis. It's very recognizable symbol and history date back to the nineteenth century to when Henry Dunant, a Swiss businessman, witnessed the suffering of thousands of casualties on both sides at the Battle of Solferino in 1859 who were left to die due to lack of care. Today, the Red Cross stretches its aid and resources across the entire globe ("The Beginning of the Red Cross Movement," 2017). DERA was founded in 1962 and now has volunteers and organizations at various stages of disaster readiness and management. DERA's members represent national governments, nonprofit associations, educational institutions, businesses and corporations, researchers, and more. Some members within DERA represent the United States, the United Kingdom, Canada, Brazil, Japan, Germany, New Zealand, and many more ("DERA: The International Association for Disaster Preparedness and Response," 2015). The World Health Organisation (WHO) was established in 1948 with more than 7000 employees and 150 country offices, with a headquarters in Geneva. While the WHO focuses on an enormous number of other issues, one of great importance is that of climate and health. The WHO explains that while humans have and will continue to have the ability to adapt to climate change, there are still very dangerous limits to our capabilities to adapt to extreme changes. Some extreme changes happen over a short period of time, such as extreme cold or hot, heavy rains, and flooding. Others are more gradual, such as a buildup of pollutants from stagnant air, or the creation of an environment suitable for mosquito population growth leading to the transmission of new or hard-to-treat diseases. The WHO states that the extreme heat of the European summer of 2003 caused 27,000 more deaths compared to the averages of previous years. Other statistical examples from past years show just how devastating climate and weather extremes can be, and include over 600,000 deaths worldwide due to weather-related disaster during the 1990s, most of which (95%) occurred in poor countries, as described throughout this book. In 2002, there were more than a staggering 3.3 million deaths across the world in relation to climate-sensitive diseases. Meteorologically, the WHO focuses on the research supporting global warming, with more extreme weather events across the world, including more frequent and stronger El Niño events, and sea level rise, projecting sea level to rise to as much as 88 cm (34.5 in) by 2100. The WHO plans to address what it sees as the most

important global topics. These topics vary, but include such things as heatwaves, variable precipitation patterns, rising sea levels, and disease and immunization control ("Climate and Health," 2005).

The United Nations (UN) is well known across the world for aspects of peace and war, economic treaties and sanctions, democracy, health, human rights, but also for humanitarian and disaster relief assistance. The UN groups different systems and organizations together in order to respond to issues that create population displacements for various reasons, including weather and natural disasters. In fact, the General Assembly of the UN has created two internationally observed days focusing on these very issues, World Humanitarian Day on 19 August, and the International Day for Natural Disaster Reductions, which falls on the second Wednesday of October every year. The UN "works to prevent disasters whenever possible, whether natural or man-made" and "the UN is in the forefront of addressing the perils of climate change which has already begun to increase the number and intensity of 'natural' disaster situations worldwide" ("Humanitarian and Disaster Relief Assistance," n.d.).

The Global Disaster Alert and Coordination System (GDACS) is a cooperative program designed to improve worldwide alerts and the exchanging of information for the initial response to major disasters. This system includes disaster managers from around the world along with the UN and the European Commission. The GDACS provides real-time information on a Web-based service and is managed by an advisory group in Yerevan, Armenia. Governments throughout the world rely on GDACS alerts and its impact estimations in order to better plan and respond to disasters locally and abroad. There are over 14,000 subscribers to GDACS, again many of which include disaster managers at the governmental level, with nations even including GDACS as a required tool when responding to and planning for disasters ("About GDACS," 2014).

15.3.1 **Global Disaster Management Technology**

World organizations and education, research, and governmental cooperation have led to an abundance of avenues and projects aimed at better management before, during, and after extreme weather- and climate-related disasters. The technology that has been introduced to the world to help with disaster management has played its part as well in order to help protect life, livelihoods, and property. As discussed in the previous chapter, drone technology has been on the rise, with about 41 countries using drone-like technology in 2004 increasing to 76 countries by 2011, and it is much higher today. The influence this technology has on disaster management is endless when it comes to the monitoring and assessment of damages. Eco-drones can supplement data collection with real-time ecosystem inventory and accounting. Drones can help with the mapping of river and coastline erosion, deforestation, and urban expansion. Unmanned aircraft can help mitigate the risks of weather-related disaster by mapping the areas that have been impacted, broadcasting video and still shots from a perspective not usually obtained, and by monitoring the spread of forest fire, flood coverage, and tropical cyclone or tornado destruction. Drone technology can also help preempt disasters by collecting and distributing information on landscape, population, and other features that may increase flood risk, landslide risk, and even volcanic eruption risk. Drones can even help monitor evacuations and show where weather conditions are worsening, which can help rescue efforts and emergency response (Harriman and Muhlhausen, 2013, pp. 7–9).

Since cellphones and smartphones have virtually become part of the human body, they have become important tools for disaster management. With sensor technology being introduced for smartphones, especially in areas where weather data are sparse, better forecasts of and preparation for impending weather events will greatly reduce the adverse effects of disasters.

Besides that, their connection to social media, text messaging, and emergency communications have and continue to be invaluable. Being able to put first-hand accounts of disaster areas or extreme weather events help the world see what people are experiencing. It supports the call for more help and aid, and it helps family and friends stay connected during events when in the past they could have become separated, adding to the chaos and anxiety of a disaster situation. Since 2009, Australia has used its Emergency Alert over 330 times in over seven million messages to alert people within areas of warning and risk by means of landline and mobile phones. Some of these warnings include alerts for fires, tsunamis, storm surge, and flooding ("Technology to Manage Natural Disasters and Catastrophes," 2012).

Due to the extreme economic risk that large-scale disasters potentially pose for countries and regions, a lot of attention has been placed on developing and deploying updated and various information and communication technologies (ICTs). Countries across the world are working to integrate newer ICTs to streamline information across different disaster management agencies. Countries, including Australia, Canada, Colombia, France, India, Indonesia, Italy, Japan, Mexico, Turkey, and the United States, are also increasing their communication technologies to enhance warnings and monitoring power of seismic and tsunami surveillance units ("Technology to Manage Natural Disasters and Catastrophes," 2012). The Group on Space Technologies for Disaster Management (STDM) focuses on promoting the awareness on how space technologies can help in identifying and managing imminent disaster events. The main objective of the STDM is "to inform the general public about how space-derived information [is] valuable in disaster management, to provide channels for new perspectives for research efforts used in space disaster management and an interdisciplinary forum to those who have an interest in disaster management and space activities, and their impact on society" ("Group on Space Technologies for Disaster Management (STDM)," 2016). With project leaders, Meshack Kinyua, a Kenyan graduate in space sciences, and Sinead O'Sullivan, an aerospace engineer from the United Kingdom, the STDM has been holding conferences around the world, including in China, Ghana, France, and the United States ("Group on Space Technologies for Disaster Management (STDM)," 2016).

Other various technologies used to combat disaster events include first-response technologies, Web-based planning tools, directories of victims, and Web-based reporting forums. The first responder Web-based application is used to track emergency responders in real time. This application can show the turn-by-turn routing and locations of responders, and includes various information for responders, such as the locations of fire hydrants and other potential hazards. InaSAFE is another Web-based tool used to help with decisions made in the aftermath of natural disasters. This tool was first used in Jakarta, Indonesia and can show on maps where specific roads may be blocked, where schools or hospitals may be closed, and helps to accurately show how many and what types of supplies are required to support the given population during the given disaster. Virtual Assembly Point (VAP) software is a directory used to help transmit the locations and physical status of the victims of extreme events. This is a text messaging service that is displayed on a Web application used by an emergency operation center which then coordinates the information and communicates it to loved ones and aid workers. This software was created and customized by the Kenyan Red Cross Society. An Australian-created Web-based application called Resilience is considered a "community-based disaster resilience in a box" and helps communities come together and report conditions, resolve non-life-threatening issues, and with the help of the Open311 upgrade even allows the Australian government to extract the reports and deliver a response when needed ("Which Technology to Use for Disaster Management?" 2013).

15.3.2 Weather-Related Insurance

One other aspect to the increased awareness and occurrence of extreme-weather- and climate-related disaster is the need for individuals to protect themselves from the financial losses that may occur. While different regions of the world have different insurance requirements when it comes to homes, businesses, vehicles, boats, agriculture, etc., it is evident that weather-related loss is creating the next huge wave of insurance policies available that are sometimes forced on the public. In some countries, such as Costa Rica, insurance policies are extremely expensive; therefore, they are not promoted to private farmers or companies. The large cost of insurance is due to the country's lack of an accurate assessment of value and possible risk to private assets (Izumi and Shaw, 2015, p. 114). Italy's insurance companies play a much more passive role when it comes to insurance policies, as they do not provide insurance contracts in areas of high natural hazard risks, while in Switzerland there is an 86% support rating for the governing building insurance and the risk transfer mechanism (Izumi and Shaw, 2015, p. 78). Since 2010, 21% of new homes built in London are in high-risk areas for flooding ("Flood Insurance: Waves of Problems," 2014). Across all of Europe, flood insurance differs in scope and form. There is no scheme or type of insurance that could be implemented across the entire continent. Each country within Europe has different risks, or issues, which promotes the need for multiple styles of flood insurance. One area, however, that can be made more universal is with policies that enhance the information of flood risk and flood-risk assessment. As is the case in most areas where flood insurance is offered, concerns about affordability and availability are often discussed. A possible solution to those concerns in Europe, and across the world, is in fact to focus on flood prevention by putting more money into the public funding of flood risk management in the form of prevention instead of insurance subsidies (Surminski, *et al.*, 2015, pp. 24–25). The United States has a very expensive and specific weather-related insurance system, where certain coastal locations have to include hurricane insurance in the homeowner's policies, and other locations in flood zones must include flood insurance. This does help when more likely situations occur, leaving the homeowner covered, but many issues occur when homes are just outside these zones and homeowners choose not to buy the expensive policies but later end up being unfortunate victims of nature's wrath. Regular home insurance policies most of the time will not cover damage caused by a flood or hurricane, leaving the homeowner responsible for all costs. Basically, those in specific zones are forced to pay expensive policies for disasters that may not occur, but those outside of the high-risk zones who choose not to pay for the expensive policies are held responsible should any damage occur. In theory this makes sense, but the issue then becomes the overly expensive rates, and the inaccurate assessments of possible risk zones. Due to the increased variability of the climate, many locations around the globe may be faced with risks that are unknown and erratic in nature, increasing the need for a much more improved assessment at local levels (Izumi and Shaw, 2015, p. 131). Building codes which are often developed by local governments are also showing a greater involvement in everyday life, especially in areas prone to tropical cyclones, tornadoes, and even in colder environments in regards to insulation. See Appendix Figure H for the overall and insured losses for loss events worldwide from 1980 to 2015.

15.4 Global Radar and Satellite Cooperation

Radar and satellite technology is still expanding, even after an explosion of advancements over the past few decades. Radar technology, be it dual-polarization radar in more and more locations or phased array technology in the United States, has shown itself to be the most important tool to a forecaster when looking at real-time up-to-the-second storm reports. Satellites, on the other hand, are quite possibly the most important tool for a forecaster from short- to long-range forecasting, as well as having its own value in nowcasting and even the overall understanding of the atmospheric makeup of our planet. It is these values that these two tools have that make them invaluable pieces for solving the meteorological puzzle. Without these technologies, meteorologists would essentially be working blind. And this is why it is vital to expand these technologies to cover the world.

15.4.1 Global Radar Effort

While many nations have found it difficult to universalize radar systems, one plan currently in place to address the need for a more global radar system is the Group on Earth Observations (GEO) high-frequency radar (HFR) system. This radar system is not a land-based radar typically used in day-to-day forecasting. Instead, this system measures ocean surface currents to support the monitoring of marine and coastal areas. This system would be used in ocean forecasting, adding to the data often missed or misrepresented by our current weather computer models. This HFR transmits radio signals that scatter off ocean surface waves. "The scattered signals are Doppler shifted by the ocean wave velocity as well as an underlying ocean current velocity. Once the velocity due to the surface gravity wave is removed, then the radial current toward or away from the radar can be measured. Combining the radial measurements of currents from several stations provides a map of the 2-D structure of the surface current" (Roarty, *et al.*, 2014). The global effort of this system was actually first put into motion in England, in 2012. The goals set forth for this network were to increase the total amount of coastal radars around the world, standardize the format of the HFR data, create worldwide standards, and assimilate the data into computer models (Roarty, *et al.*, 2014). There are over 40 nations working together on this GEO HFR system.

It is a "new international network of the combined efforts of research institutions, governments and companies to create a global system" demonstrating a perfect example of global cooperation to benefit all, rather than the withholding of valuable knowledge to benefit the few ("Global HF Radar Network," 2015).

15.4.2 Global Satellite Effort

Even though many nations now have the technological capabilities and economic resources to launch and maintain their own satellites, the costs and technological burdens make it virtually impossible for a single country to do so without international cooperation. The Initial Joint Polar System (IJPS) agreement was made between the European Organization for the Exploitation of Meteorological Satellites (EUMETSAT) and the National Oceanic and Atmospheric Administration (NOAA) in regards to a low earth orbiting satellite. The EUMETSAT is responsible for the morning coverage while the NOAA is responsible for the afternoon coverage. Furthermore, an IJPS and Joint Transition Activities (JTA) agreement calls for the sharing of instruments for each of the satellites, the exchanging of all data in real time, and for overall mutual assistance. While the satellites in this agreement carry identical sensors, the NOAA is responsible for most of the joint instrumentation, and the EUMETSAT is responsible for the development of the microwave humidity sounder (MHS) ("METOP," n.d.). This single example of international cooperation and sharing provides the whole world with a more complete understanding and mapping of the atmosphere and all that goes along with it. What could not have been accomplished by one agency or nation is now not only achievable but also advancing further. Distinguished Gates Scholar director of the Applied Geomatics Research Laboratory at the University of Waterloo Dr. Su-Yin Tan states that the "development of new meteorological satellite systems creates an increasing need for improved international coordination and cooperation for data sharing, distribution, and global weather forecasting and severe weather warnings. This is necessary to guarantee timely data access for maximizing societal benefits of meteorological satellite systems" and further states that "the development of meteorological satellites and coordinated sharing of data from these networks are perhaps the most important new development in weather forecasting and monitoring of climate change" (Tan, 2014, p. 129). Figure 15.1 shows the current polar satellite programs for the United States and its partners.

The Earth Clouds, Aerosol and Radiation Explorer (EarthCARE) is another joint mission satellite program between Europe, Japan, and Canada. This satellite is scheduled for launch in 2018. The objective of this satellite is to gather data on cloud profiles, aerosols, and precipitation. The satellite is equipped with a cloud profiling Doppler radar, a high-spectral-resolution lidar, and a multispectral imager (Illingworth, *et al.*, 2015, p. 1311). It is planned to increase "our ability both to understand cloud, aerosol, and precipitation processes, while simultaneously evaluating and improving models" (Illingworth, *et al.*, 2015, p. 1328). This mission will cover the topics of climate modeling, terminal velocity rates of precipitation, better representation of cloud structures and types, precipitation and particle sizes or profiles, the classification of the different types of absorbing and non-absorbing natural and synthetic particles, and aerosol measurements, all in hopes to better initialize models, to validate information, and to understand further our knowledge of currently unknown meteorological and atmospheric processes (Illingworth, *et al.*, 2015, p. 1328).

15.5 Global Data-Processing and Forecasting Systems

Once globalized data are collected, they have to be processed and disseminated for interpretation, analysis, and forecasting. The WWW program has put in place the Global Data-Processing and Forecasting System (GDPFS). This system is designed to prepare meteorological analyses

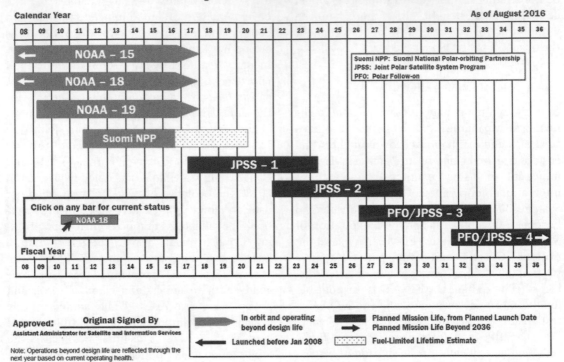

Figure 15.1 The chart shows the current polar satellite programs for the United States and its partners. Some of the partner programs include the Joint Polar Satellite System Program (JPSS) and the Suomi National Polar-orbiting Partnership, Suomi NPP. *Source*: Courtesy of NOAA Photo Library.

and forecasting products and to make them readily available, at a minimal cost to all. The organizational structure of this system has three levels: world meteorological centers (WMCs), regional specialized metrological centers (RSMCs), and national metrological centers (NMCs) (Philander, 2012, p. 1497). Each level is responsible for carrying out the functions of the GDPFS at the global level (WMCs), the regional level (RSMCs), and the national level (NMCs). In addition to these centers, the WMO and other international organizations and agencies also support the GDPFS. Some of the main functions of this system include the pre- and post-processing of NWP data and the preparation and analysis of the global 3D atmospheric

structure. The GDPFS is also responsible for the preparation of climate-related data, NWP model development, the long-term storage of data and verification results for operational and research uses, and the maintenance of updated data and products ("The Global Data-Processing and Forecasting System," n.d.). This system unifies the world as one, sharing all information, and spreading the knowledge and tools needed to help create a more standardized level of forecasting quality and accessibility.

15.5.1 Global NWP Growth

The future of NWP is a great unknown. What is known is where some advancements are needed. One of which is QPF, or quantitative precipitation forecasting. This still is an issue with all modeling; although it has improved, there is still too much error in QPF. The previously discussed convective-permitting models have greatly addressed this issue but still have room to improve. Other areas of predicted growth include the development of non-hydrostatic global deterministic NWP with resolutions of less than 5 km (3 miles), global ensembles with resolutions of less than 20 km (12 miles), embedded convective scale ensembles, sophisticated data assimilation systems, and global and convective models and ensembles that have an accurate depiction of relevant land and ocean processes. While all of these areas of improvement would and will improve accuracy and reliability, there are still other areas requiring further research. Agencies and organizations around the world are working together on trying to accomplish all these goals in order to improve NWP ("Anticipated Advances in Numerical Weather Prediction (NWP), and the Growing Technology Gap in Weather Forecasting," n.d., pp. 14–16).

15.5.2 Global Supercomputers

In order to be able to fulfill all of the goals in regards to developing NWP in the future, the issue of computer power must be addressed. As high-resolution modeling becomes the new fad in modeling, weather agencies have to balance the desire to increase modeling power by increasing their supercomputers that run the model. The higher the resolution, the greater the power necessary or the longer the computing time will be. A general guide for modeling resolution and computer power is that for every increase in model power by a factor of two the computing power needed to support such growth needs to be 10 times as much. If computing power does not keep up with modeling growth of the same rate then the time it takes to run the model will be 10 times longer ("Climate Modeling," 2012). Supercomputer company Cray builds the most used computer in mete gencies. More than 60% of the world's weather

centers use Cray computers, including but not limited to the United Kingdom, the United States, Germany, and Switzerland (Lerman, 2015). A large percentage of weather agencies around the world utilize some of the latest advancements in supercomputer technology. Fulfilling the goals of advancements in modeling on supercomputers will only progress the accuracy and reliability of weather and climatological forecasting.

15.6 Global Response to Climate Change

With the world coming together to achieve technological growth in a variety of areas in the field of meteorology, the world is also coming together in regards to one of the most pressing, extreme, and controversial issues: climate change. Climate change issues blanket the entire globe. Weather events or climate shifts of one region have a ripple effect and therefore influence the weather and climate of other regions. Changes in one location will always create changes in another. When that second location experiences its own climate change, that in turn creates changes in a third location, and so on and so forth, until the entire globe is experiencing climate changes.

To respond to the rippling effect climate change threat, numerous agencies, projects, and programs have been developed and implemented throughout the world. One such organization in the United States is the United States Agency for International Development (USAID). This agency is partnered with the US State Department, NASA, the US Environmental Protection Agency, the NOAA, and others, and strives to put countries across the globe on a path toward clean energy and low carbon development. The USAID works in about 50 countries, spread across much of Central and South America, Africa, the Indian subcontinent, Southeast and Central Asia, and Europe. By sharing data, knowledge, and state-of-the-art tools and technology, the USAID is helping countries predict and prepare for climate change The USAID has projects to reduce greenhouse gasses and to help people, governments, and institutions in developing countries to adapt with and help slow climate change. Whether it is through forest fire fighting education, land development education, installation of water catchment structures in drought-prone locations, marine sanctuaries, or climate-smart agriculture, the USAID plays a pivotal role in global climate change aid and research ("Global Climate Change Initiative," 2016).

The Met Office also has its fair share of international projects, including the Global Climate Observing System (GCOS) and the WCRP. The GCOS is an operational system developed to monitor earth's climate by detecting and attributing climate change. Co-sponsored by the WMO, the International Council for Science (ICSU), the International Oceanographic Commission (IOC), and the United Nations Environment Programme (UNEP), the GCOS helps validate climate models as well as detect levels of greenhouse gases in the atmosphere. The WCRP is also a joint program between the WMO, ICSU, and IOC. The purpose of the WCRP is to predict both climate change and how human activity may affect the climate. Since this program is a large international program, the WCRP has the ability to focus on aspects of the climate that many nations or smaller agencies are not able or capable to. The WCRP can look at the climate as the global issue that it is instead of a solely regional or national issue. The WCRP also greatly contributes to some of the more major observational and modeling studies ("International Projects," 2015).

Some other worldwide agencies stretch from the United States, to Europe, Asia, Africa, and South America. A handful of these organizations are the United Nations Office for Disaster Risk Reduction (UNISDR), the Africa Climate Change Resilience Alliance (ACCRA), the World Health Organization (WHO), the Climate Action Network (CAN), and the Intergovernmental Panel on Climate Change (IPCC). Some even

have the backing and support from some of the world's most highly regarded agencies. Two of the more prominent climate change organizations, the GCOS and the international CLIVAR project (Climate and Ocean: Variability, Predictability, and Change), have the support of and sponsorship from the EUMETSAT, NASA, the NOAA, the WCRP, and the ESA (European Space Agency).

One of the more major issues in regards to climate change is the threat that it has on the polar ice caps and, in turn, sea level rise and coastal flooding. The WWRP Polar Prediction Project was set up to focus on that very issue. This project aims to "promote cooperative international research enabling the development of improved weather and environmental prediction services for the polar regions on time scales from hours to seasonal" ("The World Weather Research Programme: A 10-year Vision," 2015). It will achieve this by improving the data assimilation and initialization of the Polar Regions in models, by investigating the stable boundary layers and sea-ice dynamics, by improving environmental predictions in the Polar Regions and the polar observing systems, and to coordinate additional observations to support modeling and verification. The Year of Polar Prediction (YOPP), with a main time period of mid-2017 through mid-2019, is said to be at the center of a major international effort to "obtain greatly enhanced polar observations and prediction capabilities" and is made up of four major elements, including "an intensive observing period, a complementary intensive modeling and forecasting period, a period of enhanced monitoring of forecast use in decision making, including verification and a special educational effort" ("The World Weather Research Programme: A 10-year Vision," 2015).

The Nordic Conference of Climate Change Adaption allows scientists, practitioners, and stakeholders to meet and discuss research, experiences, plans, and practices in regards to climate change and adaptation to those changes. There have been three conferences prior to 2016, with a fourth schedule for mid-2016.

This specific conference is of great importance as it is the first conference to be held after the historic Paris Agreement reached in 2015 (Sylte, 2016) (Figure 15.2).

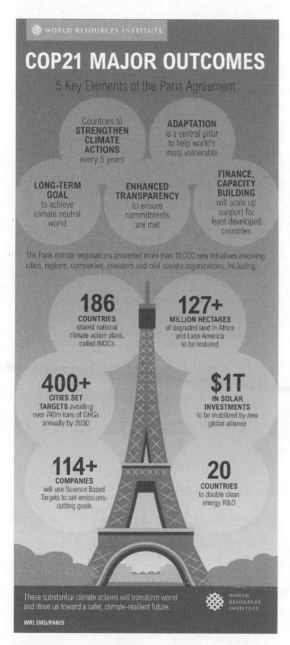

Figure 15.2 The major aspects of the historic Paris Agreement. *Source*: Reproduced with permission of World Resources Institute.

The Paris Agreement was the first ever universal and legally binding global climate deal in history. This deal is considered one of the most significant global deals of any kind, and was adopted by 195 total countries. Designed to take effect in 2020, the agreement outlines a plan to keep earth from warming to dangerous levels, limiting total warming to below 2°C and to aim to limit the increase to 1.5°C. Also agreed to is the meeting of governments every five years to keep up with science and set more ambitious goals, and to report on the progress of their own country. This will allow them to be transparent and track progress, in order to strengthen society as a whole to deal with climate change impacts, and to support developing countries. Governments are also required to enhance the understanding of early-warning systems, risk insurance, and emergency preparedness ("Paris Agreement," 2016). This agreement sets the tone for the future of meteorology and climate change. The 195 countries came together because they realized that a global agreement was necessary in order to make a difference. Individual nations may always strive to limit greenhouse gas emissions but it would never make a difference if a neighboring nation set no limits. Instead, the Paris Agreement put climate change on its own pedestal. It greatly portrayed the significance, danger, strength, and world-encompassing reach of climate change. The Paris Agreement set a standard for all to live by in order to help limit the changes to the atmosphere that, it is argued, are the root cause of many of the extreme weather events affecting the planet.

Section Summary

A common theme of this chapter is global cooperation. The coming together and the pooling of resources – of research, of technology, and financially – is key to the next phase in the development of meteorology. Whether due to climate change or extreme weather, through disaster management and technology growth, radar, satellite, and NWP advancements, or increase in supercomputer technology, there are still areas of weakness and areas in need of improvement. However, with the most recent advances, and the planned advances to come, there are signs of improvement. State-of-the-art technology has pushed the envelope of meteorology. Whether it involves tropical cyclone forecasting, coastal flooding monitoring, extreme weather modeling, nowcasting advancements, or radar and satellite technologies, never has the field touched such limits before. Accompany that growth with the development of disaster management systems, both locally and globally, the education and research in regards to climate change and extreme weather, the improvement in weather forecasting, modeling, and technologies, and in overall communicating technologies and systems, the world is heading in the right direction (Izumi and Shaw, 2015, p. 136). These advancements can only occur if nations work together to fight the common foe.

References

Book References

Izumi, T. and Shaw, R. (2015) *Disaster Management and Private Sectors: Challenges and Potentials*, Springer, Tokyo.

Philander, S. (2012) *Encyclopedia of Global Warming & Climate Change*, Second edition. SAGE Publications, Inc., New York.

Rogers, D. and Tsirkunov, V. (2013) *Weather and Climate Resilience: Effective Preparedness through National Meteorological and Hydrological Services*, The International Bank for Reconstruction and Development/ World Bank, Washington DC.

Tan, S. (2014) *Meteorological Satellite Systems*, Springer, New York.

World Bank (2008) *Weather and Climate Services in Europe and Central Asia: A Regional Review*, International Bank for Reconstruction and Development/World Bank, Washington DC.

Journal/Report References

"Anticipated Advances in Numerical Weather Prediction (NWP), and the Growing Technology Gap in Weather Forecasting." (n.d.) *WMO, with contributions from the CBS/DPFS Chairperson*, https://www.google.com/url?sa=t&rct=j&q=&esrc=s&source=web&cd=4&cad=rja&uact=8&ved=0ahUKEwjW2dXj7sbLAhVIdz4KHeuCCicQFggzMAM&url=https%3A%2 F%2Fwww.wmo.int%2Fpages%2Fprog%2Fwww%2Fswfdp%2FMeetings%2Fdocuments%2 FAdvances_NWP.pdf&usg=AFQjCNEgJKTGtt8f0JXaLlqPXdg6hpJelw&sig2=xATfMi0gjakTVnk4873gDg 20 March 2016.

Brasseur, G., and Carlson, D. (2015) Future Directions for the World Climate Research Programme. *Eos*, Climate Change, 30 July 2015.

Harriman, L. and Muhlhausen, J. (2013) A New Eye in the Sky: Eco Drones. *UNEP Global Environment Alert (GEAS)*.

Illingworth, A., Barker, H., Beljaars, A., *et al.* (2015) The EarthCARE Satellite: The Next Step Forward in Global Measurements of Clouds, Aerosols, Precipitation, and Radiation, *BAMS*, 96 (8).

Raoult, B., Thole, C., and Gartel-Zafiris, U. (2015) *Grid Technology Makes Weather Forecasts without Boundaries a Reality*. ERCIM News: Online Edition, ERCIM News 70, http://ercim-news.ercim.eu/en70/rd/grid-technology-makes-weather-forecasts-without-boundaries-a-reality. 27 March 2016.

Roarty, H., Hazard, L., Wyatt, L., *et al.* (2014) The Global High Frequency Radar, Coastal Ocean Observation Laboratory – Rutgers University, *Earthzine*, 30 October 2014.

Surminski, S., Aerts, J., Botzen, W., *et al.* (2014) Reflection on the Current Debate on How to Link Flood Insurance and Disaster Risk reduction in the European Union, *Centre for Climate Change Economics and Policy*, Working Paper No. 184, *Grantham Research Institute on Climate Change and the Environment*, Working Paper No. 162.

"The Human Cost of Weather Related Disasters (1995–2015)." (2015) *ReliefWeb*. 23 November 2015, http://reliefweb.int/report/world/human-cost-weather-related-disasters-1995-2015. 27 March 2016.

Website References

"About GDACS." (2014) *GDACS: Global Disaster Alert and Coordination System*, European Union, http://portal.gdacs.org/about, accessed 28 March 2016.

"About Us." (n.d.) *Pacific Climate Change Science*, Australian Government, http://www.pacificclimatechangescience.org/about-us, accessed 27 March 2016.

"Climate and Health." (2005) *Climate Change and Human Health*, WHO, http://www.who.int/globalchange/news/fsclimandhealth/en, accessed 28 March 2016 2016.

"Climate Modeling." (2012). *UCAR Center for Science Education*, http://scied.ucar.edu/longcontent/climate-modeling, accessed 1 April 2016.

"Collaborative Projects." (n.d.) *Collaboration for Australian Weather and Climate Research*, Australian Government Bureau of Meteorology & CSIRO, http://www.cawcr.gov.au/research, accessed 27 March 2016.

"DERA: The International Association for Disaster Preparedness and Response." (2015) *DERA*, http://www.disasters.org/. 27 March 2016.

"Earth & Environmental Sciences Journals." (2016) OMICS International, http://www.omicsonline.org/earth-and-environmental-sciences-journals.php, accessed 27 March 2016.

"EUMETNET and EU: About Us." (n.d.) *EUMETNET*, http://www.eumetnet.eu/about-us, accessed 27 March 2016

"Flood Insurance: Waves of Problems." (2014) *The Economist*, Finance and Economics, 8 March 2014, http://www.economist.com/news/finance-and-economics/21598664-new-proposals-reform-subsidised-flood-insurance-do-too-little-reduce, accessed 5 April 2016.

"Global Climate Change Initiative." (2016) *USAID: From the American*, People, https://www.usaid.gov/climate, accessed 23 March 2016.

"Global HF Radar Network." (2015) *Global HF Radar*, http://rucool.marine.rutgers.edu/geohfr/index.html, accessed 30 March 2016.

"Group on Space Technologies for Disaster Management (STDM)." (2016) *Space Technologies for Disaster Management*, Space Generation Advisory Council, http://spacegeneration.org/projects/space-technologies-for-disaster-mgmt.html, accessed 29 March 2016.

"Humanitarian and Disaster Relief Assistance." (n.d.) *Global Issues*, United Nations, http://www.un.org/en/globalissues/humanitarian, accessed 28 March 2016.

"International Projects." (2015) *Met Office*, http://www.metoffice.gov.uk/about-us/what/international/projects, accessed 1 April 2016.

"International Workshop on Disaster Risk Reduction and Management." (2015) *Strategic International Research Cooperative Program*, JST, http://www.jst.go.jp/sicp/ws2015_j-rapid_result.html, accessed 27 March 2016.

"J-RAPID." (2012) *Strategic International Research Cooperative Program*, JST, http://www.jst.go.jp/inter/english/sicp/country/j-rapid.html, accessed 27 March 2016.

Lerman, R. (2015) *Supercomputer-maker Advances Weather Forecasts*, The Seattle Times, 1 October 2015, http://m.phys.org/news/2015-10-supercomputer-maker-advances-weather.html, accessed 1 April 2016.

"METOP." (n.d.) *Monitoring Weather and Climate from Space*, EUMETSAT, http://www.eumetsat.int/website/home/Satellites/CurrentSatellites/Metop/index.html, accessed 31 March 2016.

"Natural Hazards." (n.d.) *WMO*, https://www.wmo.int/pages/themes/hazards/index_en.html, accessed 27 March 2016.

"Paris Agreement." (2016) *Climate Action*, European Commission, http://ec.europa.eu/clima/policies/international/negotiations/paris/index_en.htm, accessed 30 March 2016.

"Pilot Projects and Related Initiatives." (n.d.) *WMO*, http://www.wmo.int/pages/prog/www/ WIS/pilot_projects_related_initiatives_en. html#Simdat_VGISC_Reg6, accessed 27 March 2016.

"Sustainable Development Goals." (2016) *Sustainable Development Knowledge Platform*, United Nations, Department of Economic and Social Affairs, https:// sustainabledevelopment.un.org/?menu=1300, accessed 27 March 2016.

Sylte, G. (2016) *From Research to Actions and Transformation*, Bjerknessenteret. Facsimile of the 2nd Circular for the 4th Nordic Conference on Climate Change Adaption, http://www.bjerknes.uib.no/en/article/ news/research-transformation, accessed 1 April 2016.

"Technology to Manage Natural Disasters and Catastrophes." (2012) *OECD – Better Policies for Better Lives*, http://www.oecd.org/sti/ outlook/e-outlook/stipolicyprofiles/ newchallenges/technologytomanage naturaldisastersandcatastrophes.htm, accessed 28 March 2016.

"The Beginning of the Red Cross Movement." (2017) *British Red Cross*, http://www.redcross. org.uk/About-us/Who-we-are/History-and- origin/Beginning-of-the-Movement, accessed 24 January 2017.

"The Global Data-Processing and Forecasting System." (n.d.) *World Weather Watch Programme*, WMO, http://www.wmo.int/ pages/prog/www/DPS/gdps.html, accessed 28 March 2016.

"The World Weather Research Programme: A 10-year Vision." (2015) *WMO*, Weather, 3 March 2015, 64 (1), http://public.wmo.int/en/ resources/bulletin/world-weather-research- programme-10-year-vision-0, accessed 14 April 2016.

"Which Technology to Use for Disaster Management?" (2013). *IRIN: The Inside Story on Emergencies*, http://www.irinnews. org/analysis/2013/05/30/which-technology- use-disaster-management, accessed 1 April 2016.

Figure References – In Order of Appearance

"Mission Overview." (n.d.) GOES-R. NOAA & NASA, http://www.goes-r.gov/mission/ mission.html, accessed 17 October 2016.

Waskow, D., and Morgan, J., (2015) The Paris Agreement: Turning Point for a Climate Solution. World Resources Institute. Image updated 2016, http://www.wri.org/ blog/2015/12/paris-agreement-turning-point- climate-solution, accessed 18 July 2016.

Book Summary

This book covers the entire timeline of weather forecasting. We began with 3000 BC and ended with current and future issues of meteorology and climate change, on both a national and global scale. We have seen that weather was an unpredictable monster for thousands of years and only within the past 150 years has meteorology been explored in greater depths. Within this time, weather forecasting became practical, and more specifically in the past 50 years there has been significant accuracy advancements in weather forecasting. If you break down the time that mankind has walked on the earth, you can see that 50 years is a remarkably small speck of time. However, the advancements that occurred within the past 150 years and especially the past 50 years have saved countless lives, property, and money. This is an achievement like no other, but there is still so much more that can be done. Over a 20-year period from 1995 through 2015, statistics show that 90% of disasters worldwide were weather-related, with the United States, China, India, the Philippines, and Indonesia making up for most of these. Weather-related disasters, extreme weather events, and overall climate change are global issues. There are definitely local, regional, and national effects, but the fight is a global one. Since 1995, globally 606,000 people have died, and 4.1 billion people have been injured, displaced, or in need of some sort of emergency assistance due to weather-related events. Financially the numbers do not get any better, with an estimated $250–$300 billion annually in weather-related losses ("The Human Cost of Weather Related Disasters (1995–2015)," 2015). These numbers are astronomical, but this is the battle we face today and in the future when it comes to climate change and extreme weather. What wasn't known 5000 years ago was learned over time and through grueling research and inventions. There is no reason why that same path will not continue. Science will continue to evolve, and meteorology is no exception. Advancements in technology and a better understanding of atmospheric and oceanic processes today will make for a better prepared tomorrow. Global cooperation with the sharing of data

and research is essential at this point in time. Using different methods and modeling has its advantages in allowing for different views and ideologies; however, the compiling of data and combining of ideas and results is needed in order to propel meteorology forward. Imagine what could come of a global joint effort in gaining knowledge about weather and climate change, and where we could be in another 50 years from now. We will leave you with one final thought: "Weather has a daily impact on Earth, its environment, and biosphere. Through continued global scientific research, in depth understandings of the atmosphere, and adhering to evolution and adaptation, life forms and the environment can continue to thrive on Earth, despite weather and climatological changes".

Appendix Figures A–H

Chapter Eight

NWP Weather Models

Appendix Figure A The chart presents some of the world's more powerful computer models as of June of 2013. The left-hand column is the center and country, the second column is the name of the numerical weather prediction (NWP) system, the third is the region of coverage, the fourth column is the resolution of the model, the fifth column represents the distance out in time that the model can run, and the last column on the right is the total number of vertical layers inside the model. (Note that there are now more models running in agencies around the world, and some of the presented models are now in higher resolutions as computer power has continued to grow rapidly and the desire and need for higher resolution models has become a priority). For any updated information, please see https://www.wmo.int/pages/prog/www/swfdp/Meetings/documents/Advances_NWP.pdf, under related files. Image and data courtesy of the Data Processing and Forecasting Systems Group at WMO. Source: Adapted with permission of WMO.

Center, Country	NWP System	Domain	Horizontal Resolution	Maximum Lead-time	Vertical Levels
BoM, Australia	ACCESS-G (UM)	Global	~40 km N320	10 days	70
	ACCESS-R (UM)	Regional	~12 km	3 days	70
	ACCESS-C (UM)	Brisbane, Perth, Adelaide, VICTAS, Sydney	~4 km	36 hours	70
	ACCESS-TC (UM)	Tropical Cyclone – Re-locatable	~12 km	3 days	50
	ACCESS-Coupled Climate Model	Global	Atmos: ~250 km T47 Ocean: ~200 km, enhanced Tropics	1–9 months	38 50
CMA, China	GFS	Global	~30 km T639	10 days	60
	GEPS	Global	~60 km T213	10 days	31
	GRAPES	Regional	15 km	3 days	31
	REPS	Regional	15 km	60 hours	31
	Typhoon Det & EPS	Global – Re-locatable	~60 km T213	5 days	31
	AGCM	Global	Atmos: ~200 km T63 Ocean: ~200 km	1–6 months	16 30
CMC, Canada	GDPS	Global	~25 km	10 days (15 days on Sundays)	80
	GEPS	Global	~66 km	16 days	74
	RDPS	Regional	10 km	54 hours	80
	REPS	Regional	~33 km	3 days	28
	HRDPS	North America Canada regions	10 km 2.5 km	24 hours	80 58
	CanSIPS	Global	Fully coupled with Ocean	1 month–1 year	40
	GEM-MACH15 (air qual.)	Regional	10 km	2 days	58

Organization	Model	Domain	Resolution	Forecast range	Levels
CPTEC, Brazil	AGCM	Global	~45 km	7 days	64
	AGCM-EPS	Global	~100 km	15 days	28
	AGCM-MRF	Global	~200 km	6 months	28
	BRAMS	Regional	5 km	84 hours	50
	BRAMS-CCATT	Regional	25 km	3 days	38
	ETA	Regional	15 km	7 days	50
		Southeast Brazil	5 km	3 days	50
		Northeast Brazil	10 km	3 days	50
	ETA-EPS	South America	40 km	11 days	38
		South Brazil	5 km	3 days	50
	OA-GCM	Global	2-tier ~200 km	6 months	28
	ETA-LRF	Regional	40 km	6 months	38
DWD, Germany	GME	Global	20 km	174 hours	60
	COSMO-EU	Regional	7 km	78 hours	40
	COSMO-DE	Germany	2.8 km	21 hours	50
	COSMO-DE-EPS	Germany	2.8 km	21 hours	50
ECMWF, Europe	IFS-HRES	Global (coupled to ocean wave model)	Atmos: ~16 km T1279 Ocean waves: ~28 km (~10 km European waters)	10 days (5 days)	91
	IFS-ENS	Global (coupled to ocean wave model); 51 members	~32 km T639 Ocean waves: ~55 km	10 days	62
		Global (coupled to ocean wave model and ocean model); 51 members	Atmos: ~64 km T319 Ocean: 0.3–1 degree Ocean waves: ~55 km	10–32 days	62 42
	IFS-SEAS	Global (coupled to ocean wave model and ocean model); 51 members	Atmos: ~80 km T255 Ocean: 0.3–1 degree Ocean Waves: ~111 km	7 months 13 months (4 times/year)	62 42

(Continued)

Appendix Figure A (Continued)

Center, Country	NWP System	Domain	Horizontal Resolution	Maximum Lead-time	Vertical Levels
IMD/ NCMRWF, India	GFS	Global	~23 km T574	10 days	64
	GEPS	Global	~75 km T190	10 days	28
	UM (non-hydrostatic)	Regional 2 domains: 30 degree E–125 degree E; 9 degree S–50 degree N;76 degree E–9 degree E; 26 degree N–29 degree N	12 km 4 km	10 days	70
	WRF	North Indian Ocean India Indian regions	27 km 9 km 3 km	3 days	38
JMA, Japan	GF-S monsoon	Global	~40 km T62	4 months	38
	GSM	Global	~20 km T959	9 days (12 UTC init.)	60
	One-week EPS (WEPS)	Global	~55 km T319	9 days (12 UTC init.)	60
	One-MONTH EPS	Global	~110 km T159	34 days (Once a week)	60
	MSM	Japan and its surrounding (East Asia)	5 km	15 hours (00061218 UTC init.) 33 hours (03,09, 15, 21 UTC init.)	50
	LFM	Eastern part of Japan	2 km	9 hours (8 times a day)	60
	Typhoon EPS (TEPS)	Global	~55 km T319	5.5 hours	60
	Seasonal EPS	Global – Coupled	Atmos: ~180 km T95 Ocean: 0.3–1.0 degree × 1.0 degree	7 months (Once a month)	40 50

Centre	Model	Domain	Resolution	Forecast length	Levels
KMA, Republic of Korea	GDAPS (UM)	Global	~25 km N512	252 hours	70
	UM-EPS	Global	~40 km N320	10 days	70
	RDAPS (UM)	Regional	12 km 4 km	3 days	70
	WRF	Regional	10 km	3 days	40
	UM-Korea	Korea	1.5 km	12 hours	70
	DBAR (Typhoon model)	Re-locatable	35 km	3 days	42
	GDAPS-LRF	Global	T106	6 months	21
MF, France	ARPEGE-IFS	Global	T798C2.4 (10.5–60 km)	102 hours	70
	PEARP	Global	T538 var mesh 2.4 (15–90 km)	108 hours	65
	ALADIN	France Tropics	7.5 km 8 km	54 hours	70
	AROME – non-hydrostatic	France Tropics	2.5 km	30 hours	60
	ARPEGE – Climate	Global	Atmos: T127 Ocean: 0.5–1 degree	6 months	31 31
	MOCAGE 3D (air qual.)				
NOAA/NCEP, USA	GFS	Global	~27 km T574 (0–8 days) ~70 km T190 (8–16 days)	16 days	64
	GEFS	Global	~55 km T254 (0–8 days) ~70 km T190 (8–16 days)	16 days	42
	NAM	Regional USA regions	12 km 4 km	84 hours 2 days	60 35
	SREF (NAMB/WRF)	Regional USA regions	~16 km	87 hours	35
	Hurricane	Pacific, Atlantic	~3 km	5 days	42
	CFS	Global	Atmos: ~100 km T126 Ocean: 1/4 degree	9 months	64 40

(Continued)

Appendix Figure A (Continued)

Center, Country	NWP System	Domain	Horizontal Resolution	Maximum Lead-time	Vertical Levels
ROSHIDROMET, Russia Federation	SLAV-2008	Global	~75 km	10 days	28
	GSM	Global	~75 km (T169)	10 days	31
	BGM-EPS	Global	T169	10 days	31
			T85	30 days	30
	REG	Regional, 2 Domains: Europe + Western Siberia Eastern Siberia + Far East of Russia	~40 km ~40 km	48 hours	
	COSMO-Ru	Regional, 4 Domains: European (incl. Ural + West Siberia) part of Russia Central Russia West Russia Ural and Siberia	7 km 2.2 km 2.2 km 14 km	78 hours 24 hours 42 hours 78 hours	40 50 50 40
UKMO, UK	BGMLRF	Global	~75 km	Season	28
	GM (UM)	Global	~25 km N512	6 days	70
	MOGREPS-G-EPS (UM)	Global	~33 km N400	3 days	70
	MOGREPS-15-EPS (UM)	Global	~60 km N216	15 days	70
	UKV (UM)	UK	1.5 km (inner domain)	36 hours	70
	EURO4	Europe	4 km	5 days	70
	MOGREPS-UK-EPS (UM)	UK	2.2 km (inner domain)	36 hours	70
	AQUM (Air qual.)	UK	12 km 4 km	5 days	38
	HADGEM3-EPS (GloSea4)	Global	Atmos: ~14 km N96 Ocean: ~110 km	6 months	85 75

Figure Reference

"Anticipated Advances in Numerical Weather
   Prediction (NWP), and the Growing
   Technology Gap in Weather Forecasting."
   (2013) *WMO*, Submitted by WMO, with
   contributions from the CBS/DPFS chairperson,
   United Kingdom, pp. 20–23.

Chapter Nine

Appendix Figure B In addition to the map of the worldwide radar system of the WMO, in Chapter 9, Section 9.4, this table lists the total radar systems per country/territory, totaling over 900 radar systems, included planned radar and recently removed radar systems. Source: World Meteorological Organization.

Radar Per Member of the WMO

No.	Country	Active	Passive	Removed	Planned	Total	%
Total		877	11	14	9	911	100
1	Argentina	6	4	–	1	11	1.21
2	Armenia	3	–	–	–	3	0.33
3	Australia	56	4	6	–	66	7.24
4	Austria	5	–	–	–	5	0.55
5	Azerbaijan	2	–	–	–	2	0.22
6	Bahamas	1	–	–	–	1	0.11
7	Bangladesh	5	–	–	–	5	0.55
8	Barbados	1	–	–	–	1	0.11
9	Belarus	3	–	–	–	3	0.33
10	Belgium	3	–	–	–	3	0.33
11	Belize	2	–	–	–	2	0.22
12	Bermuda	1	–	–	–	1	0.11
13	Brazil	44	–	–	–	44	4.83
14	Brunei Darussalam	1	–	–	–	1	0.11
15	Bulgaria	3	–	–	–	3	0.33
16	Canada	31	–	–	–	31	3.4
17	Cayman Islands	1	–	–	–	1	0.11
18	Croatia	2	–	–	–	2	0.22
19	Cuba	8	–	–	–	8	0.88
20	Curaçao	1	–	–	–	1	0.11
21	Cyprus	–	1	–	–	1	0.11

(*Continued*)

Appendix Figure B (Continued)

No.	Country	Active	Passive	Removed	Planned	Total	%
Total		877	11	14	9	911	100
22	Czech Republic	2	–	–	–	2	0.22
23	Denmark	4	–	–	–	4	0.44
24	Dominican Republic	1	–	–	–	1	0.11
25	El Salvador	6	–	–	–	6	0.66
26	Estonia	2	–	–	–	2	0.22
27	Finland	10	–	–	–	10	1.1
28	France	27	–	–	–	27	2.96
29	French Polynesia	3	–	–	–	3	0.33
30	Georgia	–	–	–	1	1	0.11
31	Germany	17	–	6	–	23	2.52
32	Greece	4	–	–	–	4	0.44
33	Guyana	1	–	–	–	1	0.11
34	Hong Kong, China	6	–	–	–	6	0.66
35	Hungary	4	–	–	–	4	0.44
36	Iceland	4	–	–	–	4	0.44
37	Indonesia	34	–	–	–	34	3.73
38	Iran, Islamic Republic of	5	–	–	–	5	0.55
39	Ireland	2	–	–	–	2	0.22
40	Israel	1	–	–	–	1	0.11
41	Italy	22	–	–	–	22	2.41
42	Jamaica	1	–	–	–	1	0.11
43	Japan	29	–	–	–	29	3.18
44	Jordan	1	–	–	–	1	0.11
45	Kenya	2	–	–	–	2	0.22
46	Latvia	1	–	–	–	1	0.11
47	Macao, China	1	–	–	–	1	0.11
48	Malaysia	12	–	–	–	12	1.32
49	Mali	3	–	–	–	3	0.33
50	Mexico	13	–	–	–	13	1.43
51	Morocco	1	–	–	–	1	0.11
52	Myanmar	1	–	–	–	1	0.11
53	Netherlands (the)	2	–	–	–	2	0.22
54	New Zealand	9	–	–	–	9	0.99
55	Norway	10	–	1	–	11	1.21
56	Pakistan	7	–	–	–	7	0.77

Appendix Figure B (Continued)

No.	Country	Active	Passive	Removed	Planned	Total	%
Total		877	11	14	9	911	100
57	Panama	1	–	–	–	1	0.11
58	Paraguay	1	–	–	–	1	0.11
59	Poland	8	–	–	–	8	0.88
60	Portugal	2	–	–	–	2	0.22
61	Republic of Korea	10	–	–	–	10	1.1
62	Republic of Moldova	9	–	–	–	9	0.99
63	Romania	7	–	–	–	7	0.77
64	Russian Federation	37	–	1	–	38	4.17
65	Saudi Arabia	1	–	–	–	1	0.11
66	Serbia	14	–	–	–	14	1.54
67	Singapore	1	–	–	–	1	0.11
68	Saint Maarten	1	–	–	–	1	0.11
69	Slovakia	2	–	–	–	2	0.22
70	Slovenia	1	–	–	–	1	0.11
71	South Africa	14	–	–	2	16	1.76
72	Spain	15	–	–	–	15	1.65
73	Sweden	12	–	–	–	12	1.32
74	Switzerland	4	–	–	1	5	0.55
75	Taiwan, Prov. of China	1	–	–	–	1	0.11
76	Tajikistan	4	–	–	–	4	0.44
77	Thailand	23	–	–	–	23	2.52
78	The former Yugoslav Rep. of Macedonia	2	–	–	–	2	0.22
79	Trinidad and Tobago	1	–	–	–	1	0.11
80	Tunisia	1	–	–	–	1	0.11
81	Turkey	16	–	–	2	18	1.98
82	Ukraine	9	–	–	–	9	0.99
83	United Arab Emirates	8	–	–	–	8	0.88
84	U.K. of Great Britain and Northern Ireland	15	1	–	–	16	1.76
85	U.S.A.	220	–	–	–	220	24.2
86	Uzbekistan	3	–	–	–	3	0.33
87	Venezuela, Bolivarian Rep. of	5	1	–	2	8	0.88
88	Viet Nam	2	–	–	–	2	0.22
Total		**877**	**11**	**14**	**9**	**911**	**100**

Figure Reference

"Number of Countries Radar." (2015) *WMO Radar Database*. Operated by Turkish Meteorological Service (TMS), http://wrd.mgm.gov.tr/statistics/countries.aspx?l=en. 15 September 2015.

Chapter Twelve

NatCatSERVICE

Loss events worldwide 1980–2015
Number of relevant events by peril

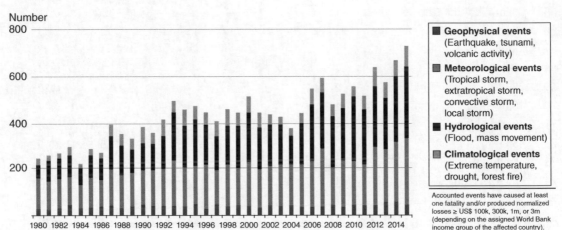

Appendix Figure C This image demonstrates the total number of relevant events by peril from 1980 to 2015. The total number of events causing death has increased steadily, especially over the past decade. Meteorological and hydrological events are among the greatest contributors to loss of life. *Source*: Reproduced with permission of Munich Re (NatCatSERVICE). (*See color plate section for the color representation of this figure.*)

Figure Reference

"NatCatSERVICE Loss events worldwide 1980 – 2015". (2016) *Munich RE*. © 2016 Münchener Rückversicherungs-Gesellschaft, Geo Risks Research, NatCatSERVICE, Munich Re, NatCatSERVICE – 2016. 18 July 2016.

NatCatSERVICE

Loss events worldwide 1980–2015
Number of severe catastrophes by peril

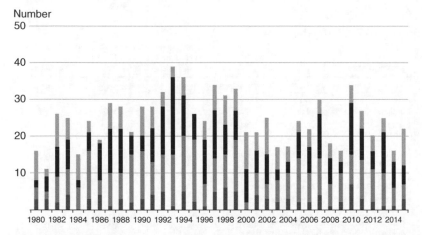

Number
50

40

30

20

10

1980 1982 1984 1986 1988 1990 1992 1994 1996 1998 2000 2002 2004 2006 2008 2010 2012 2014

■ **Geophysical events**
(Earthquake, tsunami,
volcanic activity)

▨ **Meteorological events**
(Tropical storm,
extratropical storm,
convective storm,
local storm)

■ **Hydrological events**
(Flood, mass movement)

▨ **Climatological events**
(Extreme temperature,
drought, forest fire)

Accounted events have caused ≥1,000
fatalities and/or produced normalized
losses ≥ US$ 100m, 300m, 1bn, or 3bn
(depending on the assigned World Bank
income group of the affected country).

Appendix Figure D While events causing at least one loss of life has grown steadily, severe catastrophes (events causing 1000 or more deaths) have decreased recently, especially when compared to the 1990s. This may be due to increased technology and forecasting power leading to better warnings, better forecasting and predictions, and better relief efforts and planning. This image also shows how most recent events causing severe loss of life are mostly attributed to climatological events, such as extreme temperatures, droughts, or forest fires. *Source*: Reproduced with permission of Munich Re (NatCatSERVICE). (*See color plate section for the color representation of this figure.*)

NatCatSERVICE

Significant loss events worldwide 1980 – 2015

10 deadliest events

Date	Event	Affected area	Overall losses in US$ m original values	Insured losses in US$ m original values	Fatalities
26.12.2004	Earthquake, tsunami	Sri Lanka, Indonesia, Thailand, India, Bangladesh, Myanmar, Maldives, Malaysia	10,000	1,000	220,000
12.1.2010	Earthquake	Haiti: Port-au-Prince, Petionville, Jacmel, Carrefour, Leogane, Petit Goave, Gressier	8,000	200	159,000
2–5.5.2008	Cyclone Nargis, storm surge	Myanmar: Ayeyawaddy, Yangon, Bugalay, Rangun, Irrawaddy, Bago, Karen, Mon, Laputta, Haing Kyi	4,000		140,000
29–30.4.1991	Tropical cyclone, storm surge	Bangladesh: Gulf of Bengal, Cox's Bazar, Chittagong, Bola, Noakhali districts	3,000	100	139,000
8.10.2005	Earthquake	Pakistan, India, Afghanistan	5,200	5	88,000
12.5.2008	Earthquake	China: Sichuan, Mianyang, Beichuan, Wenchuan, Shifang, Chengdu, Guangyuan, Ngawa, Ya'an	85,000	300	84,000
July – August 2003	Heat wave, drought	France, Germany, Italy, Portugal, Romania, Spain, United Kingdom	14,000	1,100	70,000
July – September 2010	Heat wave	Russia: Moscow region, Novgorod, Ryazan, Voronezh			56,000
20.6.1990	Earthquake	Iran: Caspian Sea, Gilan province, Manjil, Rudbar, Zanjan, Safid, Qazvin	7,100	100	40,000
26.12.2003	Earthquake	Islamic Republic of Iran: Bam	500	20	26,200

Appendix Figure E From 1980 through 2015 the deadliest event was the 2004 earthquake and tsunami in Sri Lanka, Indonesia and the surrounding region, resulting in 220,000 fatalities. Also making the list for the 10 deadliest events during this period included events such as tropical cyclones and heatwaves. *Source*: Reproduced with permission of Munich Re (NatCatSERVICE).

NatCatSERVICE

Significant loss events worldwide 1980 – 2015
10 costliest events ordered by overall losses

Date	Event	Affected area	Overall losses in US$ m original values	Insured losses in US$ m original values	Fatalities
11.3.2011	Earthquake, tsunami	Japan: Aomori, Chiba, Fukushima, Ibaraki, Iwate, Miyagi, Tochigi, Tokyo, Yamagata	210,000	40,000	15,880
25–30.8.2005	Hurricane Katrina, storm surge	United States: LA, MS, AL, FL	125,000	60,500	1,720
17.1.1995	Earthquake	Japan: Hyogo, Kobe, Osaka, Kyoto	100,000	3,000	6,430
12.5.2008	Earthquake	China: Sichuan, Mianyang, Beichuan, Wenchuan, Shifang, Chengdu, Guangyuan, Ngawa, Ya'an	85,000	300	84,000
23–31.10.2012	Hurricane Sandy, storm surge	Bahamas, Cuba, Dominican Republic, Haiti, Jamaica, Puerto Rico, United States, Canada	68,500	29,500	210
17.1.1994	Earthquake	United States: Northridge, Los Angeles, San Fernando Valley, Ventura	44,000	15,300	61
1.8–15.11.2011	Floods, landslides	Thailand: Phichit, Nakhon Sawan, Phra Nakhon Si Ayuttaya, Phthumthani, Nonthaburi, Bangkok	43,000	16,000	813
6–14.9.2008	Hurricane Ike	United States, Cuba, Haiti, Dominican Republic, Turks and Caicos Islands, Bahamas	38,000	18,500	170
27.2.2010	Earthquake, tsunami	Chile: Concepción, Metropolitana, Rancagua, Talca, Temuco, Valparaiso	30,000	8,000	520
23./24./27.10.2004	Earthquake	Japan: Honshu, Niigata, Ojiya, Tokyo, Nagaoka, Yamakoshi	28,000	760	46

Appendix Figure F From 1980 through 2015 the most costly events in overall losses include earthquakes, hurricanes, floods, landslides, and tsunamis. The most costly event was the 2011 earthquake and tsunami in Japan, also resulting in 15,880 fatalities. *Source*: Reproduced with permission of Munich Re (NatCatSERVICE).

NatCatSERVICE

Natural loss events worldwide 2015

Geographical overview

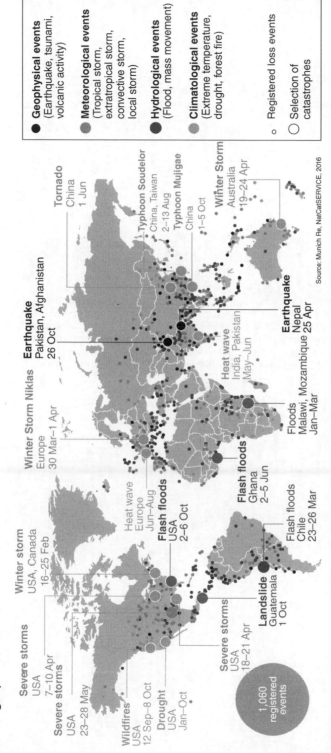

Geophysical events
(Earthquake, tsunami, volcanic activity)

Meteorological events
(Tropical storm, extratropical storm, convective storm, local storm)

Hydrological events
(Flood, mass movement)

Climatological events
(Extreme temperature, drought, forest fire)

○ Registered loss events

◯ Selection of catastrophes

Source: Munich Re, NatCatSERVICE, 2016

Severe storms
USA
7–10 Apr

Severe storms
USA
23–28 May

Wildfires
USA
12 Sep–8 Oct

Drought
USA
Jan–Oct

Severe storms
USA
18–21 Apr

Landslide
Guatemala
1 Oct

Flash floods
Chile
23–26 Mar

Winter storm
USA, Canada
16–25 Feb

Heat wave
Europe
Jun–Aug

Flash floods
USA
2–6 Oct

Flash floods
Ghana
2–5 Jun

Floods
Malawi, Mozambique
Jan–Mar

Heat wave
India, Pakistan
May–Jun

Earthquake
Nepal
25 Apr

Winter Storm Niklas
Europe
30 Mar–1 Apr

Earthquake
Pakistan, Afghanistan
26 Oct

Tornado
China
1 Jun

Typhoon Soudelor
China, Taiwan
2–13 Aug

Typhoon Mujigae
China
1–5 Oct

Winter Storm
Australia
19–24 Apr

1,060 registered events

@2016 Münchener Rückversicherungs-Gesellschaft, Geo Risks Research, NatCatSERVICE – As at March 2016

Appendix Figure G This image shows the natural loss events worldwide in 2015, including 1060 registered events, either geophysical, meteorological, hydrological, or climatological in nature. This image shows that no area is safe to extreme weather or conditions. Some of the more devastating events are highlighted with dates of occurrence. (*See color plate section for the color representation of this figure.*)

Chapter Fifteen

NatCatSERVICE

Loss events worldwide 1980–2015
Overall and insured losses

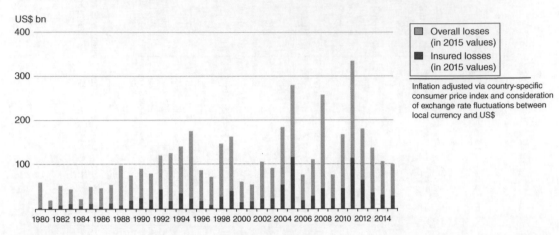

Appendix Figure H The image reflects the growing global issue of home and property insurance quality. Data show that from 1980 to 2015 overall losses due to extreme events worldwide have greatly out valued total insured losses. While there have been years that have seen greater losses, the overall trend shows overall losses more than double the values actually covered by insurance. *Source*: Reproduced with the permission of Munich Re (NatCatSERVICE).

Index

The Evolution of Meteorology: A Look into the Past, Present, and Future of Weather Forecasting, First Edition.
Kevin Anthony Teague and Nicole Gallicchio.
© 2017 John Wiley & Sons Ltd. Published 2017 by John Wiley & Sons Ltd.